Climate Change: Our Evolving Planet

Climate Change: Our Evolving Planet

Editor: Dustin Doyle

www.callistoreference.com

Callisto Reference,
118-35 Queens Blvd., Suite 400,
Forest Hills, NY 11375, USA

Visit us on the World Wide Web at:
www.callistoreference.com

ISBN: 978-1-64116-736-9 (Hardback)

Cataloging-in-publication Data

Climate change : our evolving planet / edited by Dustin Doyle.
     p. cm.
Includes bibliographical references and index.
ISBN 978-1-64116-736-9
1. Climatic changes. 2. Climate change mitigation. 3. Climatology. I. Doyle, Dustin.
QC903 .C55 2023
577.22--dc23

TABLE OF CONTENTS

Permissions

List of Contributors

Index

Preface

Climate change is the long-term ongoing shift in temperatures and weather patterns on Earth, due to human activities. The combustion of fossil fuels including gas, coal and oil emits greenhouse gases such as carbon dioxide. Greenhouse gases trap heat on Earth and do not allow it to escape, causing the greenhouse effect. Furthermore, increasing temperatures continue to melt ice sheets, which contribute in rising sea levels. This further endangers coastal and island communities, whose livelihoods are jeopardized due to increased storms, which is another result of climate change. Rising temperatures cause more water to evaporate and precipitate resulting in severe flooding. This book explores all the important aspects of climate change in the present day scenario. It is appropriate for students seeking detailed information in this area as well as for experts.

This book is the end result of constructive efforts and intensive research done by experts in this field. The aim of this book is to enlighten the readers with recent information in this area of research. The information provided in this profound book would serve as a valuable reference to students and researchers in this field.

At the end, I would like to thank all the authors for devoting their precious time and providing their valuable contribution to this book. I would also like to express my gratitude to my fellow colleagues who encouraged me throughout the process.

Editor

Rural Farmers' Approach to Drought Adaptation: Lessons from Crop Farmers in Ghana

Hillary Dumba, Jones Abrefa Danquah and Ari Pappinen

Contents

Abstract

Sub-Saharan Africa is considered to be highly vulnerable to climate change-related disasters particularly drought. Farmers in Ghana have learnt to co-exist with it by resorting to various approaches. This study sheds light on farmers' adaptation to drought in Ghana. The cross-sectional survey design was used to

H. Dumba
Institute of Education, College of Education Studies, University of Cape Coast, Cape Coast, Ghana

J. A. Danquah (✉)
Department of Geography and Regional Planning, Faculty of Social Sciences, College of Humanities and Legal Studies, University of Cape Coast, Cape Coast, Ghana
e-mail: jones.danquah@ucc.edu.gh

A. Pappinen
School of Forest Sciences, Faculty of Science and Forestry, University of Eastern Finland, Joensuu, Finland

collect data from a random sample of 326 farmers and six purposively selected lead farmers from six farming communities. Questionnaire and in-depth interviews were used for data collection. The data were analyzed using descriptive and inferential statistics. The study revealed a significant variation between locations and use of drought adaptation approaches. The study showed that the most common drought adaptation measures comprise locating farms on riverine areas, drought monitoring, formation of farm-based organizations for dissemination of climate information, application of agro-chemicals, changing planting dates, cultivating different crops, integrating crop and livestock production, changing the location of crops, diversifying from farm to non-farm income-generating activities, and cultivation of early maturing crops. Therefore, it was recommended, among other things, that Non-Governmental Organizations (NGOs) should assist the government to construct small-scale irrigation facilities and provide drought-resistant crops to further boost the capacity of farming communities in Ghana.

Keywords

Rural Communities · Subsistence Farmers · Drought · Adaptive Capacity

Introduction

Climate change has occurred and still occurring. Among all climate change-induced disasters, drought is the costliest and most devastating climatic disaster that imposes untold adverse consequences on human activities. Its recurrent occurrence is associated with high level of vulnerability among farming households (Makoka 2008; United Nations 2010). It severely affects agriculture in rural areas as well as trade and food security in both developed and developing economies of the world. Drought is particularly hazardous to communities which depend on agriculture for livelihood (Diaz et al. 2016). Incidence of drought is prevalent in Ghana, with the 1983 being the severest and most destructive in the history of the country (Owusu and Waylen 2009). Drought conditions impose consequences on crop yield and food security (Van de Giesen et al. 2010). Previous report indicated that persistent drought conditions affected all investments in the agricultural sector in the country. Unreliable rainfall, prolonged droughts, coupled with high temperatures have severely affected sustainable agriculture in the country (Armah et al. 2011; Dietz et al. 2013). Ajzen's (1985) theory of planned behavior argues that individuals perform certain planned actions known as behaviors in response to the achievement of a target. Given the serious problems posed by drought to agriculture in Ghana, farmers practice adaptation to overcome or reduce the resultant vulnerability.

Families whose livelihoods depend on farming activities need a variety of adaptation strategies to mitigate the harmful impacts of climate change and. This will help them to maintain their livelihoods (Uddin et al. 2014). Adaptation serves as the means to mitigate a system's vulnerability to hazardous events. Adaptation reflects farmer's adaptive capacity. It is a process through which a society makes better adjustments and changes in order to adapt to an unforeseen situation in the future (Smit and Wandel 2006; United Nations Framework Convention on Climate Change (UNFCCC) 2011). Adaptation refers to the process of adjustment to the actual or expected climate, its variability, and concomitant effects (Intergovernmental Panel on Climate Change (IPCC) 2014; Quandt and Kimathi 2016). It is a means to build a system's capacity, resilience, and to adjust to the impact of climate change with the ultimate aim of reducing vulnerability. It is a process through which a society makes better adjustments and changes in order to cope with an unforeseen situation in the future (Smit and Wandel 2006). It may involve adjustments in technologies, lifestyles, infrastructure, ecosystem-based approaches, basic public health measures, and livelihood diversifications to reduce vulnerability (IPCC 2014). It may also serve as means to optimizing the potential benefits of climate change. Numerous studies have examined farmers' adaptation to climate change in different locations and contexts (Mabe et al. 2014; Obayelu et al. 2014; Shongwe et al. 2014). However, these studies are not only predominantly quantitative but also based broadly on farmers' adaptation to climate change. Farmers' adaptation to climate change is dependent upon specific climate change events and hence, may differ from one climatic event to another. The measures that farmers employ to adapt to other climate change events may differ from strategies employed to adapt to drought. Therefore, a clear understanding of farmers' adaptation to drought is desirable for designing and implementing appropriate drought adaptation strategies to enhance sustainable agriculture in Ghana. The study will expand theoretical knowledge and understanding of drought adaptation planning. Specifically, the study will shed more light on farmers' planned behavior towards drought. This will provide the necessary information and reference material for other researchers and drought management policy-makers. The study also explored only farmers' views on the use of both on-farms and off-farm measures to combat drought.

Study Areas

Three agro-ecological locations, namely, Wa West (Savannah zone), Nkoranza North (Transitional zone), and Wassa East (Forest zone) of Ghana were chosen as the sites for this study (Fig. 1). Evidence indicates that rain-fed agriculture constitutes the main livelihood activity in the selected agro-ecological locations. For instance, crop farming (96.1%) is the major activity undertaken by households in the Wassa East District while most households (97.2%) in the Wa West District are engaged in crop farming as the main economic activity. Similarly, almost all agricultural households (98.5%) in the Nkoranza North District are involved in crop farming (Ghana Statistical Service 2013, 2014).

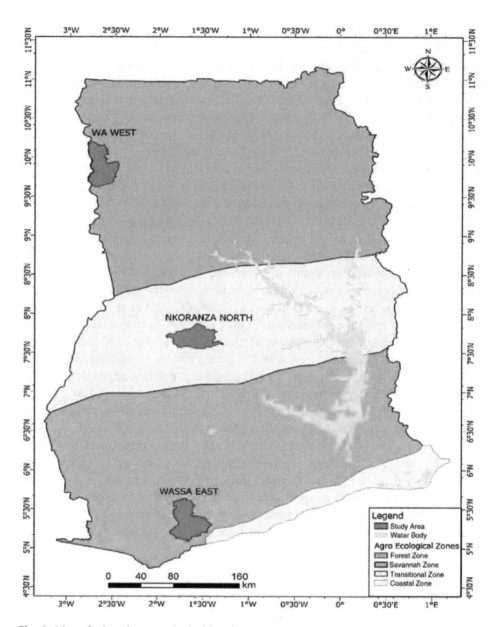

Fig. 1 Map of selected agro-ecological locations

Data Collection and Sampling Procedure

The study employed cross-sectional survey design because it has some practical advantages over longitudinal and experimental designs. Cross-sectional design helps to capture large factual numeric and descriptive data from a large sample that represents a wide target population on a one-shot basis (Bhattacherjee 2012). Out of a total population of 1765 household farmers, 326 participants were randomly selected using Yamane's (1967) formula. In generally, farming is a male-dominated activity in Ghana (Food and Agriculture Organization (FAO) 2012). However, harvesting, processing, and marketing is usually done by the female. Traditionally, the head of each household in these study areas is a male. However, in the absence of a male head the de facto female heads or de jure household heads were interviewed (See, Danquah 2015). Both qualitative and quantitative data were collected using structured questionnaire and in-depth interview guide.

Data Processing and Analysis

The first step in the process of analyzing qualitative data was data transcription. The tape recordings were listened to several times so as to get a complete sense of the data. All the transcribed data was categorized into patterns by following the guidelines prescribed by Miles and Huberman (1994) (cited in Cohen et al. 2011). Five points scale Likert estimation was used to assess or rate farmers' perception to sets of constraints to drought adaptation. The Likert scale ranging from *"Strongly Disagree (5)" to "Strongly Agree (1)"*. We also employed Pearson Chi-Square Test Statistic Tool in the analyses. This helped to compare farmers' adaptation practices across the three selected agro-ecological zones. In addition, Phi and Crammer's V were generated as measures of contingency coefficient to explore the strength of the association between the agro-ecological zones and farmers' adaptation strategies (Prematunga 2012). Problem Confrontation Index (PCI) was used as modified procedure adopted from Elias (2015) and Talukder (2014), and was computed as follows:

$$PCI = [5(P_{SA}) + 4(P_A) + 3(P_N) + 2(P_D) + (P_{SD})]$$

Where:

P_{SD} = Frequency of farmers who rated the problem as strongly disagree
P_D = Frequency of farmers who rated the problem as disagree
P_N = Frequency of farmers who rated the problem as not sure
P_A = Frequency of farmers who rated the problem as agree
P_{SA} = Frequency of farmers who rate the problem as strongly agree

Results and Discussion

Farm Household Characteristics and Adaptation Capacity

We asked participants to indicate their level of formal education, age, years of schooling, farming experience, farm size, landholding, household size, and dependents (Fig. 2). The proportion of farmers in the Forest agro-ecological zone who obtained middle school education (9.51%.) is less than the proportion of farmers in Transitional agro-ecological zone with middle school education (13.80%). This

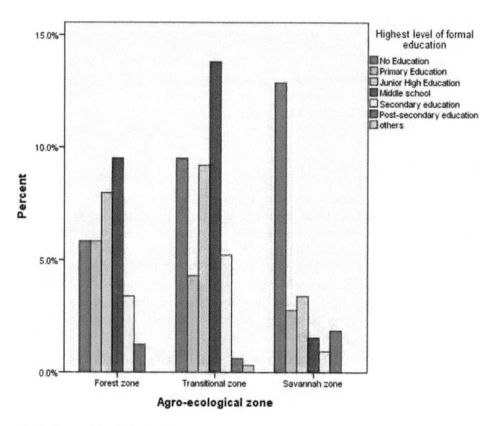

Fig. 2 Farmers' level of education across agro-ecological zones

implies that most farmers in the Forest and Transitional zones had completed middle school compared to their farming counterparts in the Savannah zone where majority (12.88%) had no education.

The educational background of farmers presupposes that these farmers would have knowledge and understanding of climatic events as well as climate change adaptation. According to Apata, (2011), education among farmers can promote climate change adaptation. Similarly, other empirical evidence from a study conducted by Abdul-Razak and Kruse (2017) indicated that farmers with formal education had high adaptive capacity while farmers without formal education had low adaptive capacity to cope with climate change and variabilities

The minimum age of the farmers was 18 years while the maximum age was 87 years (Table 1). The mean age of the farmers was 43.9 years (Mean = 43.99, SD = 14.12). Similarly, the result as shown in Table 1 is indicative that the participants have been farming for almost 19 years (Mean = 18.96, SD = 13.45). Thus, the average farming experience is 18.96 years while the minimum and maximum years of farming experience are 1 year and 76 years, respectively. Farming experience contribute to the level of knowledge on climate change adaptation and risk management (Montle and Teweldemedhin 2014). It is also clear from the results that the selected farmers had an average of 6.83 acres (Mean = 6.83, SD = 6.80).

Table 1 Descriptive statistics of farm household characteristics

Variable	N	Min	Max	Mean	SD
Age (in years)	326	18.00	87.00	43.99	14.12
Years of schooling	326	0.00	22.00	6.89	4.79
Farming experience (in years)	326	1.00	76.00	18.96	13.45
Farm size (in acres)	326	1.00	55.00	6.83	6.80
Landholding size (in acres)	326	2.00	250.00	13.77	16.66
Household size	326	1.00	25.00	6.37	3.56
Number of dependents	326	0.00	10.00	2.83	2.08

Moreover, farmers had 1.0 acre and 55.0 acres as minimum and maximum farm size, respectively. It was revealed that most of the farmers who had large farms cultivated both cash and food crops. For instance, most farmers in the Forest zone planted vast acres of cocoa whereas some farmers in the Transitional zone cultivated cashew plants on large scale. The results show that farmers' farm estate landholding ranged from at least 2.0 acres to a maximum of 250 acres, with average landholding of 13.77 acres. This implies that access to land to undertake agricultural activities may not constitute a problem to the rural farmers (see, e.g., Kassaga and Kotey 2001).

The minimum and maximum household size were 1 and 25 persons, respectively. The average household size was found to be six (Mean = 6.37, SD = 3.56) and the number of dependents in households ranged from zero to a maximum of 10. Households had about three dependents on the average (Mean = 2.83, SD = 2.08). Household size constitutes labor endowment of the farm household and it is an integration part of on-farm labor provision in smallholder farming systems (Deressa et al. 2009).

Constraints to Drought Adaptation

Farmers may have knowledge and information on drought adaptation. However, these farmers may not be capable of adapting to drought because certain factors that can hinder their adaptation behavior (see, e.g., Ajzen 1987, 2006). The results highlight that there are several challenges that confront farmers. It is evident from the results shown in Table 2 that a majority of 261 farmers (85.3%) agreed that shortage of water for irrigation is problem that confronts their capacity to cope with the impacts of drought on their farming activities. The associated PCI indicates that shortage of water for irrigation ranks first among all the problems that farmers face. This situation can be attributed to the absence of major water sources coupled with reduced precipitation in the selected agro-ecological zones (Owusu and Waylen 2009). The shortage of water poses a challenge to farmers who would have otherwise wished to irrigate their farms during episodes of drought. This confirms results of a study by Abid et al. (2015) that shortage of water for irrigation is challenge that limits farmers adaptation to drought. Furthermore, the results indicate that out of the

Table 2 Constraints to farmers adaptation to drought

Farmers' responses ($N = 326$)

Constraints to adaptation	Strongly disagree		Disagree		Not sure		Agree		Strongly agree		[a]PCI	Rank
	n	%	N	%	n	%	N	%	n	%		
Shortage of water for irrigation	15	4.6	30	9.2	3	0.9	162	49.7	116	35.6	1312	1
Unavailability of financial resources	25	7.7	38	11.0	3	0.9	127	40.0	134	41.2	1288	2
High cost of agricultural inputs	14	4.3	49	15.0	3	0.9	159	48.8	101	31.0	1262	3
Inadequate labor force	30	9.2	107	32.8	7	2.1	127	39.0	55	16.9	1048	4
Inadequate knowledge	16	4.9	143	43.9	7	2.1	112	34.4	48	14.7	1011	5
Inadequate access to extension services	45	13.8	109	33.4	3	0.9	116	35.6	53	16.3	1001	6
Inadequate time for planning	21	6.4	173	53.1	7	2.1	82	25.2	43	13.2	931	7
Inadequate access to weather information	70	21.5	131	40.2	4	1.2	68	20.9	53	16.3	881	8
Inadequate landholding	82	25.2	138	42.3	4	1.2	67	20.6	35	10.7	813	9

NB: [a]Problem Confrontation Index (PCI)

326 respondents, a majority of 85.3% farmers agreed that unavailability of financial resources serve as a constraint to their effort to adapt to drought (Table 2). This was ranked as the 2nd problem that confronts their capacity to adapt to drought. Lack of finance has been cited as the common problem that considerably hampers most farmers from adopting improved varieties of seed to combat drought (Fisher et al. 2015; Pardoe et al. 2016). High cost of agricultural inputs was rated or ranked 3rd and Inadequate labor force 4th by the farm household heads within the three ecological zones the study was conducted. However, farmers were least concern about access to land for their farming activities. This was ranked 9th on the list of factors influencing farm household ability to adapt to drought in the study area. Invariable, it was expected that extension education and information to weather forecast should feature prominently on top of the ranking, but turned out otherwise. Lack of agriculture extension services and climate base information in the form of weather forecast have been cited as the policy constraints to adaptation strategies of smallholder farmer in the tropics, particularly in sub-Saharan Africa including Ghana (Naab et al. 2019).

Determinants of Adoption of Drought Adaptation Measures

This section focuses on the presentation and discussion of main results on farmers' adoption or non-adoption of various drought adaptation measures. It also discusses farmers' socio-demographic factors as determinants of drought adaptation strategies. Table 3 presents the results with respect to farmers' drought adaptation across the three agro-ecological locations in Ghana. The application of agro-chemicals as a drought adaptation measure is significantly associated with agro-ecological locations as shown by the ($\chi^2 = 43.98$: DF $= 2$, $N = 326$), $p < 0.001$). It is indicative from the results that majority of farmers (90.7%) in Nkoranza North in the Transitional zone applied agro-chemical compared to farmers in the Daboase and Wechaiu (54.5% and 63.2%, respectively) who adapted to drought through the application of agro-chemicals. Most crop farmers in Daboase in the Transitional zone adopted the application of agro-chemicals compared to other farmers in the Forest zone because the Forest oxysol soil has higher moisture holding capacity and fertility and therefore more capable of supporting crop production. On the whole, the study reveals that most farmers (72.1%) in the selected study areas adopted application of agro-chemicals as measure to adapt to drought. This finding is consistent with results of previous studies that applying both organic and inorganic fertilizer on farmlands is a method of mitigating low crop yield associated with unreliable rainfall pattern and prolonged dry spell (Kurothe et al. 2014; Kloos and Renaud 2014; Pardoe et al. 2016).

The results as shown in Table 3 indicate that majority of farmers in Daboase in the Forest zone do not resort to migration as a drought adaptation measure. Out of the 110 farmers in the Forest zone who participated in the survey, an overwhelming majority of 101 (91.8%) did not employ migration while only nine farmers (8.2%) resorted to migration as a measure to reduce their vulnerability to drought.

Table 3 Adaptation measures across agro-ecological zones ($N = 326$)

Adaptation Measures	Agroecological Zones						Over all adoption	χ^2	ρ-value	Phi
	Forest($n = 110$)		Transitional ($n = 140$)		Savannah($n = 76$)					
	A (%)	NA (%)	A (%)	NA (%)	A (%)	NA (%)				
Application of agro-chemicals	60 (54.5)	50 (45.5)	127 (90.7)	13(9.3)	48 (63.2)	28 (36.8)	235(72.1)	43.98	0.001***	0.37
Changing planting time	52 (47.3)	58 (52.7)	119 (85.0)	21 (15.0)	63 (82.9)	13 (17.1)	234(71.8)	49.33	0.001***	0.39
Migration	9(8.2)	101 (91.8)	46 (32.9)	94 (67.1)	48 (63.2)	28 (36.8)	103(31.6)	63.04	0.001***	0.44
Cultivation of different crops	44 (40.0)	60 (60.0)	16 (11.4)	124 (88.6)	5(6.6)	71 (93.4)	261(80.1)	42.58	0.001***	0.36
Changing location of crops	68 (61.8)	42 (38.2)	21 (15.0)	119 (85.0)	62 (81.6)	14 (18.4)	223(68.4)	70.34	0.001***	0.47
Soil moisture conservation practices	12 (10.9)	98 (89.1)	47 (33.6)	93 (66.4)	28 (36.8)	48 (63.2)	87(26.7)	21.39	0.001***	0.27
Cultivation of drought-tolerant crops	16 (14.5)	94 (85.5)	21 (15.0)	119 (85.5)	16 (21.1)	60 (78.1)	53(16.3)	1.68	0.430NS	0.07
Cultivation of early maturing crops	68 (61.8)	42 (38.2)	131 (93.6)	9(6.4)	65 (85.5)	11 (14.5)	264(81.0)	41.66	0.001***	0.36

Diversifying from farm to non-farm activities	48 (43.6)	62 (56.4)	69 (49.3)	71 (50.7)	45 (59.2)	31 (40.8)	165(50.6)	4.38	110^{NS}	0.10
Integrating crop with livestock production	53 (48.2)	57 (51.8)	84 (60.0)	56 (40.0)	47 (61.8)	29 (38.2)	184(56.4)	4.68	0.970^{NS}	0.12
Home gardening	50 (45.5)	60 (54.5)	48 (34.3)	92 (65.7)	32 (42.1)	44 (57.9)	130(39.9)	3.41	0.180^{NS}	0.10
Water harvesting practices	19 (17.3)	91 (82.7)	9(6.4)	131 (93.6)	19 (25.0)	57 (75.0)	47(14.4)	14.87	0.001***	0.21
Changing size of farm land	84(60)	56(40)	41 (37.3)	69 (62.7)	36 (47.4)	40 (52.6)	161(49.4)	12.89	0.001***	0.20
Drought monitoring	41 (37.3)	69 (62.7)	91 (70.0)	42 (30.0)	45 (59.2)	31 (40.8)	184(56.4)	27.15	0.001***	0.30

NB: * Implies significant, NS implies not significant at 0.05 (2-tailed), A Adopted, NA Not Adopted

$p < 0.05*$, $p < 0.01**$, $p < 0.001***$

Moreover, out of the 140 farmers in Nkoranza North in the Transitional zone that participated in the study, it was found that a greater proportion of farmers (67.1%) did not adopt migration as drought adaptation measure. Thus, migration is not a common drought adaptation strategy in the Forest as well as the Transitional zones of Ghana. This contradicts findings of Yang et al. (2015) that migration is the commonest drought adaptation strategy among farmers in the Ningxia Hui Autonomous Region of North-western China. Most farmers in the Forest and Transitional zones of Ghana do not adapt to drought through migration because there are various livelihood options and crop diversification strategies that assist them to adapt to the hardships imposed by drought (Asante et al. 2017). For instance, various artisanal activities, trading or business ventures, seeking employment in craft and cottage industries, and other sources of off-farm income generating abound in the forest belt of Ghana and hence, most farmers in this area do not over dependent on rain-fed agriculture. However, there are more cases of migration among farmers in Wechaiu in the Transitional zone compared to farmers in Daboase in the Forest zone (see Derbile et al. 2016). This is because some farmers in the Forest zone migrated either from the Savannah zone, Transitional zone, or neighboring communities in Cote d' Ivoire to undertake cocoa cultivation since the rainfall pattern in the Forest is more favorable to farming activities (Jarawura 2013).

The results suggest that farmers in the Savannah zone are more likely to adapt to drought and rainfall variability through migration to other places compared to farmers in the Forest and Transitional zones of Ghana. This collaborate the findings Van der Geest (2011) and Jarawura (2013), that rainfall variability and climate change slightly account for the out-migration of farmers from the three northern regions to Brong Regions of Ghana. When there are drought conditions some farmers migrate to other areas to engage in other livelihood activities. There is a statistical highly significant relationship between agro-ecological zones and farmers' adoption of migration as a drought adaptation measure ($\chi^2 = 63$: DF $= 2$; $N = 326$; $p < 0.001$).

Migration among farmers is dependent on agro-ecological location. The phi value (0.44) indicates that is a positive significant moderate difference between farmers' migration patterns and agro-ecological zones. This is because the severity of drought differs from one agro-ecological zone to another (Adepetu and Berthe 2007).

The results indicated that out of the 110 farmers interviewed in Daboase in the Forest zone, 89.1% were nonadopters of soil moisture conservation practices as an adaptation strategy. However, relatively small proportion of the farmers (10.9%) in this zone adopted soil conservation practices. Similarly, nonadopters were 66.4% and 63.2% in Transitional and Savannah zones respectively. Collectively, across all the ecological zones studied out of 326 farmers interviewed only 87 farmers employed soil conservation measures. This represents a total of 26.7% farmers.

Moreover, it was revealed that only 53 (16.3%) out of the 326 farmers in the three agro-ecological zones cultivated some sort of drought-tolerant crops as drought adaptation measure. The results show that most farmers in the agro-ecological location do not cultivate crops that are drought-resistant. Only a small proportion of farmers in the various agro-ecological zones indicated that they cultivated some crops that are resistant to drought conditions. Farmers non-adoption of drought-resistant crop varieties can be attributed to the fact that most farmers in Ghana do not have access to drought-tolerant crops. This contradicts the results of previous works by Udmale et al. (2014) that rural farmers widely cultivate less water intensive and drought tolerant crops as adaptation options to drought.

The results show that a majority of 264 farmers (81.0%) out of the 326 farmers in the three agro-ecological zones adopted the cultivation of early maturing crops as a measure to adapt to drought. The adoption of early maturing crops is highly dependent upon agro-ecological zone. Most farmers in the Savannah and Transitional zones cultivate early maturing crops compared to the proportion of farmers in the Forest zone who cultivate early maturing. The results of this current study corroborate the findings of various previous studies (Bawakyillenuo et al. 2016; Pardoe et al. 2016) that farmers resort to the cultivation of early maturing crops as a climate change adaptation strategy.

The study also revealed that farmers have been adapting to drought by integrating both farming and non-farming activities as similarly found by a previous study by Balama et al. (2013). From the results out of the 326 farmers, a little over half (50.6%) diversified from farm to non-farm income generating activities in order to adapt to the impact of drought. Majority of farmers (59.2%) who diversified farm to non-farm income generating activities were located in Wechaiu in the Savannah zone.

The study further indicated that most farmers (56.4%) integrated livestock production with crop production as drought adaptation measure. This is because farmers seek solace in livestock rearing when their crops fail as a result of drought. Farmers do experience decline in crop productivity as a result of drought. Therefore, they have seen the need to engage in livestock rearing to augment their farming activities. Similarly, Balama et al. (2013) has found that local farmers in Kilombero District of Tanzania integrated crop farming into livestock production as a climate change adaptation strategy. The results in Table 3 indicate that most farmers in Wechaiu in the Savannah zone (61.8%) as well as those in the Transitional zones (60.0%) integrated livestock rearing with crop production compared to farmers in the Forest zone (48.2). This confirms results of a study by Bawakyillenuo et al. (2016) that integrating livestock rearing into crop production is common climate change adaptation method being adopted by farmers in rural Savannah zone of northern Ghana. This is because the vegetation and climatic features within the Savannah and Transitional zones are more favorable to livestock rearing compared to the Forest zone. However, this is not significant ($\chi^2 = 4.68$: DF $= 2$, $N = 326$; $p > 0.05$).

There is moderate significant association between the proportion of farmers who employed water harvesting practices and agro-ecological zone ($\chi^2 = 14.87$; DF $= 2$; $N = 326$, $p < 0.001$). A majority of farmers (93.6%) in the Transitional zone and farmers in Savannah (75.0%) employed water harvesting practices as drought adaption measure compared to number of farmers in the Forest zone (82.7%) who did not employ water harvesting practices to combat drought. Most farmers in the Savannah and Transitional zones experienced severe drought conditions and acute water shortage than farmers in the Forest zone (Armah et al. 2011). This in keeping with the fact that farmers in the Savannah and Transitional zones need to harvest rainwater and store it for domestic use and animal consumption as well. However, farmers in the Forest may have unimpeded access to riverine water supply through-out the year due to high level of precipitation (Armah et al. 2011).

The results show strong relationship between change in farm sizes and agro-ecological zones ($\chi^2 = 12.89$; DF $= 2$; $N = 326, p < 0.001$). Variation in farm size as a drought adaptation strategy is dependent upon agro-ecological location. The proportion of farmers who change their farm sizes as drought adaptation mecha-nism varies across the agro-ecological zones (sensu, Hansen et al. 2004). The settler farmers in the Transitional zone have fixed portions of land for farming, whereas the native farmers have most of their land occupied by cashew plantation. Hence, such farmers may find it difficult to increase their farm size. Farmers in the Savannah zone may not even change their farm sizes because fertile lands are limited in supply. Hence, farmers are fixated to the same parcel of land. Moreover, the farmers may find it unrewarding and time-consuming to clear new parcel of land for cultivation in the midst of unpredictable and scanty rainfalls. However, majority of farmers 84 (60.0%) in Daboase in the Forest zone stated that they changed their farm sizes in order to deal with the impacts of drought. Finally, the results indicate that out of the 140 farmers in Nkoranza North in the Transitional zone, 70.0% in this zone adopted drought monitoring as drought adaptation strategy, particularly constant listen to weather news on radio and TV stations on daily basis. During an interview in the Transitional zone, a male farmer indicated that:

> I always listen to 'weather man' on FM radio in order to know the on-set of rains before I even begin to prepare for farming. Sometimes before I go to farm, I have to listen to 'weather man' to know whether it would rain on that day or not (Male farmer, Transitional zone).

Similarly, a majority of 45 farmers (59.2%) in the Savannah zone indicated that they employed drought monitoring as a tool for preparing for impending drought condi-tions and to improve their resilience to drought vulnerability. The plurality of radio stations as well as the availability of agricultural extension officers in the study areas provide easy access to weather information. Hence, most farmers continually mon-itor weather and climatic conditions before they plant their crops. Regarding drought monitoring, a lead farmer in the Savannah zone hinted during an interview schedule that:

> We do not sow arbitrarily in this area. We usually 'study' the weather pattern to predict the arrival of rains before sowing seeds (Male lead farmer, Savannah zone).

However, majority of 69 farmers, representing 62.7% of the 110 farmers who participated in the survey in the Forest zone did not practice drought monitoring. The climatic conditions in this zone is quite conducive for agriculture. The farmers in this zone hardly experience severe drought that lasts long as compared to farmers in the Savannah and Transitional zones Moreover, the soil in the Forest zone holds moisture. Hence, farmers in this zone do not really have to monitor the rainfall pattern as farmers in the Savannah and Transitional zones would do. Overall, more than half of farmers (56.4%) practice drought monitoring. This shows that drought monitoring is mostly being practiced by farmers as a method of adapting to drought in the selected agro-ecological zones. This confirms the results of a study by Pardoe et al. (2016) that farmers "follow the rain" until they are well-convinced that the rain would not fail them before they sow their seeds. It is obvious that drought monitoring is highly statistically and significantly related to agro-ecological zones ($\chi^2 = 27.15$: DF $= 2$, $N = 326$; $p < 0.001$). This is because various agro-ecological zones have different amount of precipitation and soil moisture content to support farming activities. The phi value (0.30) indicates that there is a moderate significant relationship between drought monitoring and agro-ecological zones. Therefore, the decision of a farmer to monitor and time drought would depend upon a particular zone where he is located. Rather than employing only scientific and orthodox strategies to adapt to drought, the study also revealed that the farmers also employ prayers and supplications as means to adapt to drought conditions. They offer supplications to Him so that He would cause the rain to fall. This could be so because farmers have sociocultural perception about climate change and drought. Some farmers attribute the occurrence of climate and drought to the intention of God and other deities (Jarawura 2013). Hence, farmers combine both spiritual and scientific means to adapt to drought.

Traditionally, we usually call on deities to intercede for us to get the rains. We go round the community to pour libation asking the gods of the land to cause rains to fall. And if it rains, we thank them [gods] by making animal sacrifice. (Male farmer, savannah zone). *During droughts, we throw a challenge to the gods of the land to let it rain to prove the that they are living gods* (Male farmer, Transitional zone).

In conclusion, farmers employed both scientific and unscientific methods to adapt to drought in the selected agro-ecological locations in Ghana. The study reveals that drought adaption measures differ significantly among farmers in the Forest, Transitional and Savannah zones of Ghana. This finding is in harmony with results of various studies (Jarawura 2014; Abid et al. 2015; Bawakyillenuo et al. 2016) that climate and drought adaptation strategies are numerous and their implementation differs from place to place. This is because farmers' knowledge of drought adaption and their adaptive capacities as well as rainfall and soil properties differ from place to place. Therefore, farmers in various geographical locations would adapt to drought by adopting different mechanisms. However, the most commonly adopted drought adaptation measures comprise application of agro-chemicals, changing of planting date, cultivating different crops, integration of crop and livestock production, changing the location of crop on yearly basis, diversifying from farm to non-farm income generation activities, cultivation of early maturing crops, and drought monitoring.

Conclusions

Farmers' adaptation to drought differs across various agro-ecological locations in Ghana and they adapt to drought by employing mixed adaptation strategies. The most commonly used drought adaptation strategies include application of agro-chemicals, changing planting dates, cultivation of different crops, changing location of crops, cultivation of early maturing crops, diversification to non-farm activities, integrating crops and livestock production, as well as drought monitoring. Moreover, farmers' choice of specific drought adaptation strategies is a determinant of various factors. Farmers' ecological location acts as the major significant determinant of their adoption of all the eight drought adaptation measures. Finally, farmers with access to credit facilities and extension services are more likely to adopt farm-based drought adaptation measures and less likely to diversify to non-farming activities. Ministry of Food Agriculture (MoFA) and the National Disaster Management Organization should provide drought relief measures and safety net programs for vulnerable smallholder farmers. This also calls for the introduction and implementation of crop insurance schemes where farmers would be given the opportunity to indemnify their crops against possible loss associated with drought. As a matter of mitigating farmers' vulnerability to drought, both governmental organizations such as MoFA and National Climate Research Institute, and other non-governmental organizations should help develop, introduce, and implement affordable drought adaptation technologies in farming communities. The introduction and cultivation of drought-resistant crops, water harvesting, and conservative agriculture practices should be promoted among farmers in the country.

References

Abdul-Razak M, Kruse S (2017) The adaptive capacity of smallholder farmers to climate change in the northern region of Ghana. Clim Risk Manag 17:104–122

Abid M, Scheffran J, Schneider UA, Ashfaq M (2015) Farmers' perceptions of and adaptation strategies to climate change and their determinants: the case of Punjab province, Pakistan. Earth Syst Dyn 6:225–243

Adepetu AA, Berthe A (2007) Vulnerability of rural sahelian households to drought: options for adaptation. A final report submitted to assessments of impacts and adaptations to climate change (AIACC), Project No. AF 92. Retrieved from http://www.start.org/Projects/AIACC_Project/Final Reports/Final

Ajzen I (1985) From intentions to actions: a theory of planned behaviour. In: Kuhl J, Beckmann J (eds) Action-control: from cognition to behaviour. Springer, Heidelberg, pp 11–39

Ajzen I (1987) Attitudes, traits, and actions: dispositional prediction of behaviour in personality and social psychology. In: Berkowitz L (ed) Advances in experimental social psychology. Academic, New York, pp 1–63

Ajzen I (2006) Theory of planned behaviour. Retrieved from http://people.umass.edu/~aizen/tpb.diag.htmlnull-link

Apata TG (2011) Factors influencing the perception and choice of adaptation measures to climate change among farmers in Nigeria. Evidence from farm households in Southwest Nigeria. Environ Econ 2(4):74–83

Armah FA, Odoi JO, Yengoh GT, Obiri S, Yawson DO, Afrifa EKA (2011) Food security and climate change in drought-sensitive savannah zones of Ghana. Mitig Adapt Strateg Glob Chang 16(3):291–306

Asante WA, Acheampong E, Kyereh E, Kyereh B (2017) Farmers' perspectives on climate change manifestations in smallholder cocoa farms and shifts in cropping systems in the forest-savannah transitional zone of Ghana. Land Use Policy 66:374–381

Balama C, Augustino S, Eriksen S, Makonda FSB, Amanzi N (2013) Climate change adaptation strategies by local farmers in Kilombero District, Tanzania. Ethiop J Environ Stud Manag 6 (6):724–736

Bawakyillenuo S, Yaro JA, Teye J (2016) Exploring the autonomous adaptation strategies to climate change and climate variability in selected villages in the rural northern savannah zone of Ghana. Local Environ 21(3):361–382

Bhattacherjee A (2012) Social science research: principles, methods, and practices. University of South Florida, Tampa

Cohen L, Manion L, Morrision K (2011) Research methods in education, 6th edn. Routledge, New York

Danquah JA (2015) Analysis of factors influencing farmers' voluntary participation in reforestation programmes in Ghana. For Trees Livelihoods. https://doi.org/10.1080/14728028.2015.1025862

Derbile EK, Jarawura FX, Dombo MY (2016) Climate change, local knowledge and climate change adaptation in Ghana. In: Yaro J, Hesselberg J (eds) Adaptation to climate change and variability in rural West Africa. Springer, Cham. https://doi.org/10.1007/978-3-319-31499-0_6

Deressa TT, Hassan RM, Ringler C, Alemu T, Yesuf M (2009) Determinants of farmers' choice of adaptation methods to climate change in the Nile Basin of Ethiopia. Glob Environ Chang 19 (2):248–255

Diaz HM, Hurlbert M, Warren J (2016) Final research report of the rural communities' adaptation to drought. Canadian Plains Research Center, Regina. http://www.parc.ca/vacea/assets/PDF/rcadrepor

Dietz T, van der Geest K, Obeng F (2013) Local perceptions of development and change in Northern Ghana. In: Yaro J (ed) Rural development in Northern Ghana. Nova Science Publishers, New York, pp 17–36

Elias AZM (2015) Farmers' problem confrontation in crop diversification at southern region of Bangladesh. J Agroecol Nat Resour Manag 2(3):230–233

Fisher M, Abate T, Lunduka RW, Asnake W, Alemayehu Y, Madulu RB (2015) Drought tolerant maize for farmer adaptation to drought in sub-Saharan Africa: determinants of adoption in eastern and southern Africa. Climate Change 133(2):283–299

Food and Agricultural Organization (2012) Gender inequalities in rural employment in Ghana: an overview. Retrieved from http://www.fao.org/docrep/016/ap090e/.pdf

Ghana Statistical Service (2013) Ghana living standard survey. Ghana Statistical Service, Accra

Ghana Statistical Service (2014) 2010 population and housing census: district analytical report. Nkoranza North. Ghana Statistical Service, Accra

Hansen J, Marx S, Weber E (2004) The role of climate perceptions, expectations, and forecasts in farmer decision making: The Argentine Pampas and South Florida. Final Report of an IRI Seed Grant Project. International Research Institute for Climate Prediction (IRI), The Earth Institute at Columbia University

Intergovernmental Panel on Climate Change (IPCC) (2014) Climate change 2014: impacts, adaptation, and vulnerability. Part B: regional aspects. Contribution of working group II to the fifth assessment report of the Intergovernmental Panel on Climate Change. Cambridge University Press, Cambridge

Jarawura FX (2013) Drought and migration in northern Ghana. Doctoral thesis, University of Ghana, Legon

Jarawura FX (2014) Perceptions of drought among rural farmers in the Savelugu district in the northern Savannah of Ghana. Ghana J Geogr 6:102–120

Kassaga K, Kotey NA (2001) Land management in Ghana: building on traditional and modernity. International Institute for Environment and Development, London

Kloos J, Renaud FG (2014) Organic cotton production as an adaptation option in north west Benin. Outlook Agric 43(2):91–100

Kurothe RS, Kumar G, Singh R (2014) Effect of tillage and cropping systems on runoff, soil loss and crop yields under semiarid rain fed agriculture in India. Soil Tillage Res 140:126–134

Mabe FN, Sienso G, Donkoh S (2014) Determinants of choice of climate change adaptation strategies in northern Ghana. Res Appl Econ 6(4):75–94

Makoka D (2008) The impact of drought on household vulnerability: the case of rural Malawi. Paper presented at the 2008 United Nations University (UNU-EHS) Summer Academy on Environmental Change, Migration and Social Vulnerability. Retrieved from http://mpra.ub.uni-muenchen.de/15399

Miles MB, Huberman AM (1994) An expanded source book: qualitative data analysis, 2nd edn. Sage, London

Montle BP, Teweldemedhin MY (2014) Assessment of farmers' perceptions and the economic impact of climate change in Namibia: case study on small-scale irrigation farmers (SSIFs) of Ndonga Linena Irrigation Project. J Dev Agric Econ 6:443–454

Naab FZ, Abubakarib Z, Ahmed A (2019) The role of climate services in agricultural productivity in Ghana: the perspectives of farmers and institutions. Clim Serv 13:24–32

Obayelu OA, Adepoju AO, Idowu T (2014) Factors influencing farmers' choices of adaptation to climate change in Ekiti state, Nigeria. J Agric Environ Int Dev 108(1):3–16

Owusu K, Waylen P (2009) Trends in spatio-temporal variability in annual rainfall in Ghana (1951–2000). Weather 64(5):115–120

Pardoe J, Kloos J, Assogba NP (2016) Seasonal variability: impacts, adaptations and the sustainability challenge. In: Yaro JA, Hesselberg J (eds) Adaptation to climate change and variability in rural West Africa. Springer International Publishing, Cham, pp 41–57

Prematunga RK (2012) Correlational analysis. Aust Crit Care 25:195–199

Quandt A, Kimathi YA (2016) Adapting livelihoods to floods and droughts in arid Kenya: local perspectives and insights. Afr J Rural Dev 1(1):51–60

Shongwe P, Masuku MB, Manyatsi AB (2014) Factors influencing the choice of climate change adaptation strategies by households: a case of Mpolonjeni Area Development Programme (ADP) in Swaziland. J Agric Stud 2(1):86–98

Smit B, Wandel J (2006) Adaptation, adaptive capacity and vulnerability. Glob Environ Chang 16:282–292

Talukder Md R (2014) Food security assessment in Sylhet Haor area emphasizing homestead productivity and agricultural resource utilization. A thesis submitted to Bangabandhu Sheikh Mujibur Rahman Agricultural University in partial Fulfillment of the requirement for the degree of Master of Science in agronomy. Sheikh Mujibur Rahman Agricultural University, Bangladesh

Uddin MN, Bokelmann W, Entsminger JS (2014) Factors affecting farmers' adaptation strategies to environmental degradation and climate change effects: a farm level study in Bangladesh. Climate 2:223–241

Udmale P, Ichikawa Y, Manandhar S, Shidaira H, Kiem AS (2014) Farmers' perception of drought impacts, local adaptation and administrative mitigation measures in Maharashtra State, India. Int J Disaster Risk Reduct 10:250–269

United Nations (2010) Natural hazards, unnatural disasters: the economics of effective prevention. World Bank, Washington, DC

United Nations Framework Convention on Climate Change (UNFCCC) (2011) Assessing climate change impacts and vulnerability: making informed adaptation decisions. Retrieved from www.unfccc.int/adaptation

Van de Giesen K, Liebe J, Jung G (2010) Adapting to climate change in the Volta Basin, West Africa. Curr Sci 98(8):1033–1037

Van der Geest K (2011) The Dagara farmer at home and away. Migration, environment and development in Ghana. PhD dissertation, University of Amsterdam. African Studies Centre, Leiden

Yamane T (1967) Statistics: an introductory analysis, 2nd edn. Harper and Row, New York

Yang J, Tan C, Wang S, Wang S, Yang Y, Chen H (2015) Drought adaptation in the Ningxia Hui Autonomous Region, China: actions, planning, pathways and barriers. Sustainability 7:15029–15056

2

Agropastoralists' Climate Change Adaptation Strategy Modeling: Software and Coding Method Accuracies for Best-Worst Scaling Data

Zakou Amadou

Contents

Abstract

Investigating software and coding method accuracies are still a challenge when dealing with best-worst scaling data. Comparing various climate change policy estimates and their relative importance across different statistical packages has received little attention. In this chapter, we use best-worst scaling approach to determine agropastoralist preferences for 13 climate change adaptation policies across two popular statistical packages (R and SAS). While data were collected from 271 agropastoralists, mixed logit was used to analyze data. Results reveal that mean and standard deviation estimates for 13 climate change adaptation policies from R are higher and statistically significant than SAS estimates. Based on R estimates, prolific animal selection, vaccination, settlement, strategic

---

Z. Amadou (✉)
Faculty of Agricultural Sciences, Department of Rural Economics and Sociology, Tahoua University, Tahoua, Niger
e-mail: zakouamadou77@gmail.com

mobility, and strategic destocking are the most popular climate change adaptation policies, and more than two-third of respondents are in favor of these policies.

Keywords

Climate change adaptation policies · Best-worst scaling · Agropastoralists · Mixed logit models

---

Introduction

A myriad of climate change adaptation policies have been proposed to improve farmers and herdsmen's resilience building capacity, diversify their income and food, and increase their welfare in a changing climate. Several computation methods have been used to analyze discrete choice and best-worst datasets. While conditional logit, multinomial logit have been widely used in the academic literature, the usage of mixed logit (ML) models has been exploded because it is capable to relax (IIA) and can approximate any random utility model (McFadden and Train 2000).

A recent research has studied mixed logit models: accuracy and software choice by comparing ML across three popular packages, namely, SAS, NLOGIT-LIMDEP, and a user-written add-in module for STATA. Results indicate that the data generating process used was not well suited to evaluate the accuracy of software packages and further research is needed to determine which software is the most accurate (Chang and Lusk 2011).

Another challenge is to determine the climate change adaptation policy, and there are few studies geared at computing welfare effects of individual climate change policy. In addition, determining the relative importance of several climate change policy options can be challenging though they may have the same objective towards resilience building. For instance, prolific animal selection and changing herd composition both aim at reducing herd size and thereby enhancing agropastoralists' resilience build capacity. A third challenge facing researcher is to write an algorithm capable of solving real-life problem.

This chapter contributes to fill a knowledge gap by eliciting agropastoralists preference for alternative climate change policies and also enriching literature related to climate change and choice modeling. While data collected from choice experiment and best-worst have been well-documented, coding methods and algorithm development vary considerably from one statistical package to another.

Background on BWS

Climate change is increasingly becoming recognized as a global threat and concerted effort such as new climate change adaptation strategies should be undertaken both at local, national, and regional and global levels.

While various government climate change adaptation policies have been proposed to reinforce vulnerable farmers' resilience capacity, little is relatively known about farmers indigenous knowledge related to climate change adaptation strategies. However, not all climate change policies have had expected results on farmers' welfare, and previous studies have indicated that combined government and farmers adaptation strategies are more welfare enhancing (Tabbo et al. 2016).

Climate Change Policy Identification and Survey Design

Based on previous research and interview with agropastoralists, 13 climate change adaptation policies have been identified and included in this study. To determine the relative importance that agropastoralists place on these policies, a BWS experiment was designed (Marley and Louvriere 2005). A balanced incomplete block design (BIBD) developed by Louviere et al. (2015) has been used to determine allocation of the 13 policies for each BW question. The resulted design contains 13 BWS questions, each having four policy options. The BIBD is the most widely used design in the BWS literature because it is not only a balanced design but also an orthogonal design (Flynn and Marley 2014). Furthermore, the policies were selected to reflect the main issues and challenges recently discussed in climate change adaptation choice as compared to the food choice literature documented by Lusk and McCluskey (2018).

For each BWS question, respondent was asked to select his best and his worst climate change adaptation policies. Figure 1 listed below reports an example of the best-worst questions used in the study.

The coding method used for R is based on position for each pair of best-worst question. The value for the position can be ranged from 1 to 4. For SAS, best options will be assigned 1, worst options (-1), and 0 for non-chosen options. This shows that a scale difference may exist between the two coding methods (Table 1).

Data Analysis

The BWS approach assumes that respondents simultaneously make repeated choices by choosing the best and worst items in a given set and thereby maximizing the difference (Flynn and Marley 2014). By denoting J as number of items in each BWS

Best	Endogenous Strategies	Worst
☐	Strategic mobility	☐
☐	Transhumance	☐
☐	Settlement	☐
☐	Income generating income	☐

Fig. 1 Please select your best and worst endogenous strategies

Table 1 Climate change adaptation policy and description

Climate change policy	Description
Mutual assistance	Helping agropastoralist to rebuild his herd by donating a gift
Settlement	Combining farming and animal rearing
Strategic mobility	Mobility orientated towards researching forage and water
Strategic destocking	Reducing herd size to accommodate shortage of forage and drought
Prolific animal selection	Keeping animals capable of yielding higher meat and milk products
Sheep and cattle fattening	Increasing sheep and cattle weight by feeding on high concentrate
Vaccination	Treating animal from infectious diseases and producing safe products
Transhumance	Unidirectional movement of the herd in searching of H_2O and forage
Emergency destocking	Reducing herd size as result of disease's outbreak
Water and soil conservation activities	Aims at increasing food and forage production
Changing herd composition	Keeping only animals that have developed strong resilience
Income generating activities	Undertaking various activities capable of boosting revenue
Forage cultivation	Selecting and growing forage to meet its increasing demand

question (4 climate change adaptation policies), then $J\,(J-1)$ best-worst pairs of best-worst choices are possible.

By following this approach, our data were analyzed using random utility framework which is well rooted in microeconomics (McFadden 1974), whereby a given respondent n derives from the selected best-worst pairs in each BWS question t is the difference in utility between the j best and k worst policies.

This can be mathematically expressed:

$$U_{njt} = \mu_{jt} - \mu_{kt} + \varepsilon_{njt} \tag{1}$$

where μ is the vector of estimated importance parameters of the best and worst climate change policies (j and k respectively) relative to some policy normalized to zero for identification.

The probability that respondents choose item j as best and k as worst out of J items in BWS question t is the probability that the difference in utility of the chosen items (U_{njt} and U_{nkt}) is greater than all other $J\,(J-1)-1$ possible differences within each BWS question (Lusk and Briggeman 2009). While several econometric methods can be used to model this behavior, mixed logit is the most widely used estimation procedure, because it is flexible and can approximate any random utility model (Train 2009).

The mixed logit model and the probability that an individual n chooses j as best and k as worst can be mathematically written as follows:

$$P_{nj} = \int \mu \prod_{t=1}^{T} \frac{e^{[\mu_{njt} - \mu_{nkt}]}}{\sum_{l=1}^{J} \sum_{m=1}^{J} e^{[\mu_{nlt} - \mu_{nmt}] - J}} f(\mu_n) d\mu_n \qquad (2)$$

where $f(\mu_n)$ is the density of the importance parameters μ_n.

The share of preferences for each climate change adaptation policy can be expressed as follows:

$$\varphi_j = \frac{e^{\mu_j}}{\sum_{k=1}^{j} e^{\mu_k}} \qquad (3)$$

where φ_j is the share of preference for a given climate change adaptation policy.

Results

Data were collected via well-structured questionnaire delivered to a sample of 271 respondents. Summary statistics and variable definitions are reported in Table 2. Most of the respondents have an average of 40 years and about 76% of respondents were male and 84% of the respondents were married. About 75% of respondents were uneducated, and about 52% and 27% of our sample have an annual income below 90,000 FCFA and between 90,0000 and 180,000 FCFA, respectively. In addition, average household size was 8 persons with an average of 16 and 24 respectively for large animal and small animal. About 87% of respondents strongly believe that climate change and environment are correlated.

Table 3 reports estimates from mixed logit models for both R and SAS software. Coefficients reflect the importance of each 12 climate change adaptation polices to forage cultivation, which was normalized to zero for identification purpose. Results show that based on R software estimation, strategic mobility followed by mutual assistance, settlement, strategic destocking, prolific animal selection, sheep and cattle fattening, vaccination, transhumance, emergency destocking, water and soil

Table 2 Characteristics of surveyed respondents

Variable	Definition	Mean	Standard deviation (SD)
Age	Age in number	40	12
Genre	1 if male, 0 female	0.76	0.42
Marital status	2 if married, 0 otherwise	0.84	0.51
Education	1 uneducated, 0 other	0.25	0.22
Base: Income3	**Above 180,000 FCFA**	**0.00**	**1**
Income1	Below 90,000 FCFA	0.52	0.27
Income2	90,000–180,000 FCFA	0.27	0.52
Family size	Size in numbers	8	5
Large animal size	Size in numbers	16	18
Small animal size	Size in numbers	24	25
Climate and environment	1 if yes	0.87	0.49

Table 3 Mixed logit models for best-worst scaling data: software and coding method accuracies

R software		SAS software	
Agropastoralists' climate change strategies	Estimate	Agropastoralists' climate change strategies	Estimate
Strategic mobility	0.887*(0.088)	Settlement	0.571*(0.080)
Mutual assistance	0.587*(0.089)	Emergency destocking	0.369*(0.075)
Settlement	0.582*(0.076)	Prolific animal selection	0.347*(0.085)
Strategic destocking	0.528*(0.077)	Water and soil conservation activities	0.299*(0.068)
Prolific animal selection	0.521*(0.072)	Strategic mobility	0.264*(0.067)
Sheep and cattle fattening	0.515*(0.079)	Transhumance	0.214*(0.069)
Vaccination	0.483*(0.071)	Mutual assistance	0.172*(0.073)
Transhumance	0.407*(0.075)	Sheep and cattle fattening	0.080(0.074)
Emergency destocking	0.266*(0.074)	Income generating income	0.052(0.066)
Water and soil conservation activities	0.183*(0.075)	Vaccination	−0.059(0.066)
Changing herd composition	0.181*(0.075)	Strategic destocking	−0.142(0.066)
Income generating activities	0.048(0.079)	Changing herd composition	−0.583(0.079)
Base: Forage cultivation	0.0000	Baseline: Forage cultivation	0.000
Standard deviation estimates		**Standard deviation estimates**	
Sd.(Strategic mobility)	1.210*(0.161)	Sd.(Settlement)	1.068*(0.153)
Sd.(Mutual assistance)	1.631*(0.163)	Sd.(Emergency destocking)	1.106*(0.148)
Sd.(Settlement)	0.598*(0.191)	Sd.(Prolific animal selection)	1.430*(0.152)
Sd.(Strategic destocking)	0.814*(0.163)	Sd.(Water and soil conservation activities)	0.330(0.251)
Sd.(Prolific animal selection)	0.104(0.522)	Sd.(Strategic mobility)	0.513*(0.201)
Sd.(Sheep and cattle fattening)	0.801*(0.158)	Sd.(Transhumance)	0.549*(0.178)
Sd.(Vaccination)	0.129(0.397)	Sd.(Mutual assistance)	0.935*(0.151)
Sd.(Transhumance)	0.895*(0.155)	Sd.(Sheep and cattle fattening)	1.069*(0.148)
Sd.(Emergency destocking)	0.897(0.153)	Sd.(Income generating activities)	0.044(0.440))
Sd.(Water and soil conservation activities)	0.879*(0.151)	Sd.(Vaccination)	0.338(0.247)
Sd.(Changing herd composition)	0.912*(0.150)	Sd.(Strategic destocking)	0.223(0.374)
Sd.(Income generating activities)	1.210*(0.151)	Sd.(Changing herd composition)	0.974*(0.149)
Log likelihood	−8586.9	−8845	
Numbers of observations(N)	14,612	14,612	
Run time	11 m:22 s	7 m:43 s	

Numbers in parentheses are standard errors, * denotes mean importance of the policy which is statistically different from forage cultivation, Sd stands for standard deviations

conservation activities, and changing herd composition are the most climate change adaptation policies with a significant difference to forage cultivation. Standard deviation estimates for 12 climate change adaptation policies obtained from R software were statistically and highly significant, implying that heterogeneity is a pattern when analyzing climate policies within the population. Table 2 also reports SAS software estimates. Results reveal that settlement followed by emergency destocking, prolific animal selection, water and soil conservation activities, strategic mobility, transhumance, and mutual assistance are the most important climate change adaptation polices relative to forage cultivation. Results also suggest that standard deviation estimates for settlement, emergency destocking, prolific animal selection, water and soil conservation activities, strategic mobility, transhumance, mutual assistance, sheep and cattle fattening, and changing herd composition are statistically significant, revealing that these estimates do vary in the population and thereby confirming heterogeneity pattern. Heterogeneity pattern when analyzing data in R is higher than heterogeneity pattern with SAS.

Table 3 also shows that R and SAS generated different mean estimates. For instance, the mean estimates for strategic mobility were 0.887 and 0.264 respectively for R and SAS. Similarly, the mean estimates for mutual assistance were 0.587 and 0.172 respectively for R and SAS. The mean estimates for settlement were 0.582 and 0.571 respectively for R and SAS. The mean estimates generated by R are all statistically significant. Similarly, the standard error estimates for strategic mobility, mutual assistance are higher for R than SAS, while standard error estimates for settlement and prolific animal selection are higher for SAS than R. This implies that estimates of standard errors considerably diverge across R and SAS. The standard deviation estimates are generally higher and significant for R than SAS; implying heterogeneity pattern is highly significant for R than SAS. Consequently, R software accurately predicted heterogeneity pattern than SAS.

Table 4 reports share of preferences for R and SAS software estimates. Results show that 12.17%, 9.02%, and 8.97% of respondents viewed strategic mobility, mutual assistance, and settlement as the most desirable climate change adaptation policies, respectively. Based on SAS estimates, 11.63%, 9.51%, and 9.30% of respondents viewed settlement, emergency destocking, and prolific animal selection as the most desirable climate change adaptation policies, respectively. This indicates that share of preference for settlement estimated with SAS is higher than estimate from R. Conversely, share of preference for strategic mobility estimated with R is higher than estimate from SAS. Share of preference for mutual assistance under R is higher than that of SAS. This shows that mixed results prevail while estimating share of preferences under R and SAS. Results also indicate that SAS's algorithm converges faster than that of R.

Table 5 reports the intention to vote for or against each climate change adaptation policy. Results based on R estimation show that more than 70% of respondents would vote for the implementation of policies such as prolific animal selection (100%), vaccination (100%), settlement (84%), strategic mobility (77%), strategic destocking, and sheep and cattle fattening (74%). Forage cultivation (50%) and income generating activities (52%) had the lowest vote share among respondents.

Table 4 Share of preferences based on R and SAS estimates

	R		SAS
Strategic mobility	12.17%	Settlement	11.63%
Mutual assistance	9.02%	Emergency destocking	9.51%
Settlement	8.97%	Prolific animal selection	9.30%
Strategic destocking	8.50%	Water and soil conservation activities	8.86%
Prolific animal selection	8.44%	Strategic mobility	8.56%
Sheep and cattle fattening	8.39%	Transhumance	8.14%
Vaccination	8.13%	Mutual assistance	7.81%
Transhumance	7.53%	Sheep and cattle fattening	7.12%
Emergency destocking	6.54%	Income generating income	6.92%
Water and soil conservation activities	6.02%	Vaccination	6.20%
Changing herd composition	6.01%	Strategic destocking	5.70%
Income generating activities	5.26%	Changing herd composition	3.67%
Base: Forage cultivation	5.01%	Baseline: Forage cultivation	6.57%

Results based on SAS estimates indicate that more than 70% of respondents would vote the implementation of policies such as income generating activities (88%), water and soil conservation activities (82%), and settlement and strategic mobility (70%). Strategic destocking (26%) and changing herd composition (27%) had the lowest vote share among respondents. Results from Tables 3 and 4 further suggest that concordance between preferences for climate change adaptation polices and voting behavior does exist. Voting implementation results reveal that R and SAS estimates greatly vary.

Conclusion

A mosaic of climate change adaptation policies have been proposed to build rural household resilience capacity and diversify income and food strategies and improve their livelihood.

This chapter uses BWS approach to elicit Niger agropastoralists preferences for climate change adaptation policies. Using data collected from 271 agropastoralists, results indicate that agropastoralists have a clear preference for settlement, emergency destocking, prolific animal selection, water and soil conservation activities, strategic mobility, transhumance, and mutual assistance. While results also suggest that mean, standard error and standard deviation estimates of these preferences vary across R and SAS package, R package yielded more accurate results than SAS.

Results further suggest that share of preferences and vote for climate change adaptation policy implementation are software package dependent. Future research is to study the stability of climate change adaptation policies overtime. Lusk (2012) stated that public policy is highly interventionist and that policy preferences are more

Table 5 Agropastoral's vote for climate change adaptation policy implementation

R software			SAS software		
Climate change policy	Vote for implementation (%)	Vote against implementation (%)	Climate change policy	Vote for implementation (%)	Vote against implementation (%)
Strategic mobility	77	23	Settlement	70	30
Mutual assistance	64	36	Emergency destocking	63	37
Settlement	83	17	Prolific animal selection	60	40
Strategic destocking	74	26	Water and soil conservation activities	82	18
Prolific animal selection	100	0	Strategic mobility	70	30
Sheep and cattle fattening	74	26	Transhumance	65	35
Vaccination	100	0	Mutual assistance	57	43
Transhumance	67	33	Sheep and cattle fattening	53	47
Emergency destocking	62	38	Income generating income	88	12
Water and soil conservation activities	58	42	Vaccination	43	57
Changing herd composition	58	42	Strategic destocking	26	74
Income generating income	52	48	Changing herd composition	27	73
Forage cultivation	50	50	Forage cultivation	50	50

likely to change external shocks when climate change hit, food safety crisis occurred, and negative externalities happened.

References

Chang JB, Lusk JL (2011) Mixed logit models: accuracy and software choice. J Appl Econ 26:167–172

Flynn T, Marley T (2014) Best-worst scaling: theory and methods. In: S. Hess and A. Daly (eds), Handbook of Choice Modelling. Cheltenham, UK: Edward Elgar 178–201

Louviere JJ, Flynn TN, Marley AAJ (2015) Best-worst scaling: theory, methods and applications. Cambridge University Press, Cambridge

Lusk JL (2012) The political ideology of food. Food Policy 37(5):530–542

Lusk JL, Briggeman BC (2009) Food Values. Am J Agric Econ 91(1):184–196

Lusk JL, McCluskey J (2018) Understanding the impacts of food consumer choice and food policy outcomes. Appl Econ Perspect Policy 40(1):5–21

Marley AA, Louvriere JJ (2005) Some probabilistic models of best, worst, and best-worst choices. J Math Psychol 49(6):464–480

McFadden D (1974) Conditional logit analysis of qualitative choice behavior. Frontiers in Econometrics. New York: Zarembka

McFadden D, Train K (2000) Mixed MNL models for discrete response. J Appl Econ 15(5):447–470. https://doi.org/10.1002/1099-1255(200009/10)15:5447::AID-JAE5703.0.CO;2-1

Tabbo AM, Amadou Z, Danbaky AB (2016) Evaluating farmers' adaptation strategies to climate change: a case study of Kaou local government area, Tahoua State, Niger Republic. Jàmbá: J Disaster Risk Stud 8(3):a241. https://doi.org/10.4102/jamba.v8i3.241

Train K (2009) Discrete choice methods with simulation. Cambridge University Press, Cambridge, UK

3

Tied Ridges and Better Cotton Breeds for Climate Change Adaptation

R. Mandumbu, C. Nyawenze, J. T. Rugare, G. Nyamadzawo, C. Parwada, and H. Tibugari

Contents

R. Mandumbu (✉)
Crop Science Department, Bindura University of Science Education,
Bindura, Zimbabwe
e-mail: rmandumbu@gmail.com

C. Nyawenze
Cotton Company of Zimbabwe, Harare, Zimbabwe

J. T. Rugare
Department of Crop Science, University of Zimbabwe, Harare, Zimbabwe

G. Nyamadzawo
Department of Environmental Science, Bindura University of Science Education,
Bindura, Zimbabwe

C. Parwada
Department of Horticulture, Women's University in Africa, Harare, Zimbabwe

H. Tibugari
Department of Plant and Soil Sciences, Gwanda State University, Gwanda, Zimbabwe

Cotton Production Under Climate Change
Effects of Water Harvesting on Soil Moisture Content .
References

Abstract

Climate change and variability is already reducing agricultural productivity and opportunities for employment, pushing up food prices and affecting food availability and production of formerly adapted crop types. Such is the case in cotton production in Zimbabwe, where it was the only viable commercial crop in marginal areas. As a form of adaptation, there is need for African farmers to have a range of agricultural techniques as coping strategies and tactics to enable sustainable production of crops and deal with extreme events. Such techniques include water conservation and introduction of new adapted crop genetics to cope with the new environment. The emerging trends in climate change will force farmers to adopt new crops and varieties and forms of agricultural production technologies. The objective of this study is to determine the contribution of combining in-field water harvesting and early maturing cotton varieties in curbing drought in cotton in semiarid Zimbabwe. The results show that both water harvesting in form of planting basins significantly (P <0.05) increased boll number and branch number of cotton across all varieties. The varieties M577 and M567 out-performed the conventional varieties in early growth, branch number, and boll number. Tied contour ridges gave a significantly (P <0.05) higher moisture content in 0–5 cm and the 6–10 cm depth compared to conventional tillage. The new varieties displayed early phenological development. Despite the existence of rainfall gaps, the in-field water harvesting techniques captured enough moisture and prevented moisture losses through runoff which resulted successful flowering and fruiting in the short varieties compared to conventional tillage on conventional varieties. In this regard, water harvesting and early maturing varieties offer considerable hope for increasing crop production in arid and semiarid areas of Zimbabwe.

Keywords

Climate change · Adaptation · Water conservation · Cotton · Semiarid

---

Introduction

Africa is regarded as having climates that are among the most viable in the world on seasonal and decadal time scales (UNFCCC 2007). Floods and droughts can occur in the same area within months from each other. Of the total additional people at risk of hunger due to climate change, although already larger proportion, Africa may account for the majority by the 2080s (Fischer et al. 2002). The increase in inerratic rainfall seasons characterized by unpredictable length of seasons, high temperatures alternating floods and dry spells, and variable rainfall amounts presents new

Table 1 Agro-ecological regions and potential land use for Zimbabwe

Agro-ecological zone	Area (ha)	Area as % of country	Characteristic weather	Recommended land use
I	703,400	1.8	High rainfall (>900 mm per annum), with some precipitation throughout the year. Low temperatures	Fruits, tea, coffee macadamia nuts, intensive livestock production
II	5,861,400	15.1	Moderately high (750–1000 mm per annum) rainfall confined to summer. Severe dry spells are rare	Intensive crop and/or livestock production
III	7,287,700	19.5	Infrequent heavy rainfall leads to moderate annual recording of about 650–800 mm. Fairly severe mid-season dry spells	Marginal for maize, tobacco, and cotton production, livestock production
IV	14,782,300	36.7	Fairly low total rainfall (450–650 mm per annum). Periodic seasonal droughts, severe dry spells during rainy season	Drought resistant crops such as sorghum and pearl millet, livestock production
V	10,441,100	26.8	Low and erratic rainfall (<450 mm per annum, <650 mm in the Zambezi valley, and <600 mm in the Sabi-Limpopo Valleys). Prolonged midterm dry spells	Too dry for successful crop production without irrigation. Marginal millet, sorghum, extensive beef ranching, game ranching

challenges to the majority of the farmers in the absence of appropriate response strategy (Zimbabwe climate change response strategy, 2017). In the 2018–2019 cropping season below normal and highly erratic and patchy rainfall was recorded for the first half of the season and most crops were stressed with most being completely written off because of prolonged dryness (FEWS Net 2019).

In Zimbabwe, over 70% of Zimbabwe's employment is directly or indirectly accounted for by agriculture. The national agricultural production largely relies on rain-fed agriculture which is one of the most vulnerable sectors to climate change and variability. Adaptation to climate change will entail adjustments and changes at every level from community to national level. Communities must build their resilience including adapting appropriate technologies while making the most of traditionally and locally generated technologies and diversifying their livelihoods to cope with the current and future stresses (Fischer et al. 2002).

Zimbabwe is generally characterized by low rainfall and more than 50% of its land area falls under region IV and V which receive rainfall lower that 650 mm per season (Table 1).

Maize planted late will not give good yields, thus making maize production a less viable activity under climate change conditions. In the low-lying areas of southern

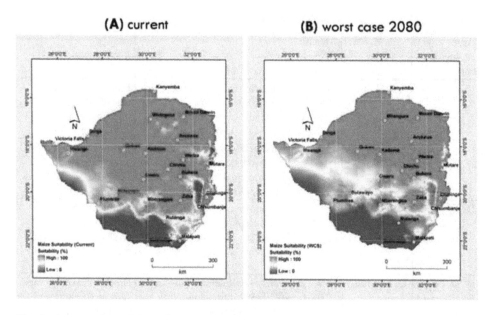

Fig. 1 A comparison of the maize production zones under the current and worst case scenario for the year 2080. (Source: Zimbabwe's Second Communication to the United Nations framework convention on climate change (UNFCCC))

Zimbabwe, for example, it is probable that climate change will turn the region into a non-maize-producing area, as exemplified by reduced maize production in Masvingo (Fig. 1). The projections are that by 2080 the area suitable for maize production in the south-western and central provinces of Zimbabwe will have decreased (Fig. 1).

This area, which represents 42% of the communal area, will become even more marginal for maize production. Based on site results, seasons could be 25% shorter than now.

Communities living in natural regions IV and V (which make up about 64% of the land area) are at the mercy of climatic extremes, with few livelihood options (Brazier 2015). They tend to be the most vulnerable to poverty. These regions are already feeling the impacts of climate change and will be the hardest hit in the future. Climate change will exacerbate hardship and poverty among the people of Zimbabwe. Women, children, and the disabled, especially those living in rural areas, will be the worst affected. The causes of rural poverty relate to the adverse climate and environmental conditions that disrupt agriculture, the main livelihood activity in areas where most people live. Women, children, the elderly and the disabled have been identified in several studies as being the most vulnerable to shocks. Zimbabwe's agricultural systems are already insecure as they depend mainly on seasonal rainfall. In addition, ruinous land use practices in the form of poor soil and water management, reduced biodiversity, and poor choice of crops to plant have led to degradation of the resource based on which agriculture depends. Climate change will hasten the degradation and exacerbate food insecurity, which is already

Table 2 The impacts, sectorial vulnerabilities, and adaptive capacity to climate change

Impacts	Sectorial vulnerability	Adaptive capacity
Warning by 1.5 °C	Agricultural production severely compromised	Rainwater harvesting
Drier subtropical regions	Uncertainty on what and when to plant	Improved varieties that fit into the current seasons
Decrease in annual rainfall	Yields from rain-fed production can be halved	
Extreme events: increase in frequency and intensity of storms and droughts and floods	Net revenue could fall by 90% by 2100	

Adapted and modified from Christensen et al. (2007)

prevalent in Zimbabwe. There are likely to be shifts in the start and end of the rainy season, and the onset of the rains may be delayed by between 4 and 6 weeks. This will mean changes in planting and harvesting dates, the length of the growing season, and the types of crops that farmers are forced to adopt.

Climate change affects crop production in the following ways (Table 2).

Characteristics of Cotton Growing Areas in Zimbabwe

The farms are generally small, often held under traditional tenure and are located in marginal areas which are risk prone (FAO 2012). Their challenge is to improve production of cotton under climate change. The investments in farming are low and their ability to adapt is very low. The inventions in this sector have to low cost so that adaptation levels are high hence the introduction of tied ridges and use of imported varieties for sustainable cotton production. Varieties purchase will be done by everybody prior to the onset of the season. Tied ridges are not an expensive technology to introduce to the farmers as long as they understand the merits.

Crop Genetic Diversity and Climate Change

The emerging changes in climate will force farmers to adopt new varieties and crop types and forms of agricultural production technologies that can respond to new and changing stress factors. The areas that are currently the most food insecure will be worst affected and will have the greatest need for new crop varieties that are tolerant to drought high temperatures flooding salinity and other environmental extremes (FAO 2015). Diverse crop species, varieties, and cultivation practices allow crops to be grown across a wide range of environments. Sometimes better adapted varieties will need to be brought in from outside.

Traits that contribute to phenotypic plasticity (ability to cope with a wide range of environmental conditions) may be increasingly important. Such is the case with cotton in Zimbabwe where the traditional varieties were complimented by imports of new cotton seed varieties from India.

Genetic resources could contribute greatly to efforts to cope with climate change. It is likely that climate change will necessitate more international exchanges of genetic resources as countries seek to obtain well-adapted crops. There is likely going to be a greater interdependence on the use of genetic resources and that underscores the importance of international cooperation (FAO 2015).

Food-insecure people in the developing world such as Zimbabwe especially women and indigenous people are among the most vulnerable groups and usually the hardest hit. In the case of cotton, the varieties used in Zimbabwe were not of a match to the current environmental characteristics. Zimbabwe farmers continued to cultivate the varieties although they were poorly adapted to the environment. The Cotton Company of Zimbabwe imported hybrid varieties from India with the objective of improving productivity and helping farmers to cope with environmental adversities.

Status of Cotton

Cotton is the second most important cash crop in Zimbabwe and is grown by thousands of smallholder farmers on average plot sizes of about 1 ha in the summer growing season (Global Agricultural Information Network 2017). The crop is strategic for poverty alleviation and is of major significance to food security for smallholder farmers in marginal areas due to its contribution to incomes and employment (Mujeyi 2013). The crop supports over one million people in marginal areas of Zimbabwe including farmers their families, farm workers, and industrial workers (Buka 2017; Mujeyi 2013). Most cotton growers have limited opportunities as these are in semiarid areas and cotton production is the only viable option.

However, cotton production has been on the decrease in terms of the number of farmers producing the crop (Fig. 2).

In some yeas the government intervened in cotton production through free cotton inputs. Generally the trend from the number of cotton farmers to yield per hectare (Figs. 2 and 3) points to the fact that cotton production is in the decrease. This might be partly due to the low prices on the international market. The low levels of production might be partly due to old varieties which are poorly adapted to the current environmental trends as dictated by climate change. There have been no new cotton varieties over the past 25 years in Zimbabwe. This implies that the recent climate shifts experienced in most parts of the world has not been factored in Zimbabwe cotton breeding program. The government of Zimbabwe prohibited the use of genetically modified seed which might have improved cotton yields.

Climate change has brought greater uncertainty and exposure to multiple climate stress. The lives of millions of people in semiarid Zimbabwe who depended on

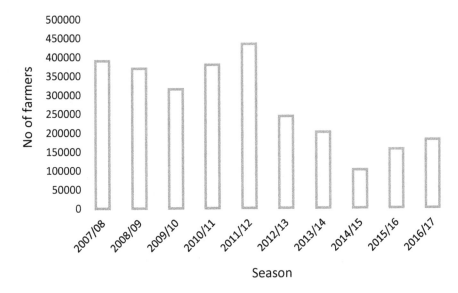

Fig. 2 Number of farmers cultivating cotton from 2007/08 season to 2016/17 season in Zimbabwe

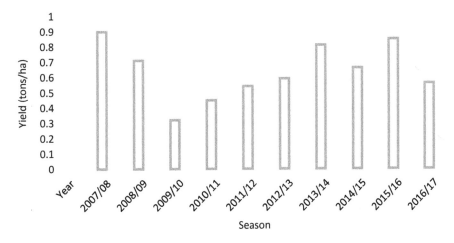

Fig. 3 Average yield of cotton (t/ha) from 2007/08 to 2016/17

cotton for their livelihood are highly vulnerable to climate change. The local farmers have traditionally managed harsh environment by growing moisture stress-tolerant cotton. In order to improve the effectiveness of crop production in these marginal rainfall regions, cultural practices which conserve water availability to the crop are essential (Mupangwa et al. 2006).

Investing in agricultural production methods to boost farmers' resilience against weather shocks is a key strategy to reduce negative impacts. The cotton growing areas of Zimbabwe are generally rain fed and are characterized by rainfall patterns which are highly variable in amount and distribution. This has been exacerbated by climate change. According to McHugh et al. (2007), the main limitation for increasing crop yields in rain-fed farming systems is crop water stress caused by inefficient use of rainwater. Rockstrom and Falkenmark (2000) reported that inefficient use of rainwater is often a consequence of poor rainfall partitioning resulting in low root-zone soil moisture or poor plant uptake of available moisture. Making rainfall available and its effective storage and efficient use are therefore important adaptive mechanisms and major determinants in cotton production under climate change. According to Nyamadzawo et al. (2013), climate change models have projected a decrease in rainfall in southern Africa and research has already shown the same. Therefore, the focus should be on upgrading rain-fed smallholder farming in tropical environments characterized by frequent droughts and mid-season dry spells. According to Ibraimo and Munguambe (2007), there is need for more efficient capture and use of scarce water resources in arid and semiarid areas. The optimization of rainfall management through water harvesting in sustainable and integrated production system can contribute to improve small scale farming households by upgrading rain-fed agricultural production. Research has documented substantial increases that are obtained through soil water conservation and efficient use of it by the crop, and, subsequently, the yield increases.

In-Field Moisture Harvesting

Tied Ridges

The effect of the ridges was generally higher in drier periods and more when the ridge ends were tied. Belay et al. (1998) reported that in wetter seasons, open-end ridges gave higher yields which show how important it is that soil and land management practices include means to safely dispose water from the field, should the rain exceed the retention of the soil. UNEP (1997) also confirmed that rain water harvesting involves the use of methods that increase the amount of water stored in the soil profile by trapping or holding the rain where it falls and it involves small movements of rainwater in order to concentrate it where it is required.

Planting Basins

In Zimbabwe, planting basins are structures that are dug from July through October in the same positions annually and whose recommended dimensions are 15 cm length by 15 cm width and 15 cm depth (Mupangwa et al. 2006; Twomlow and Hove 2007). These basins are spaced at 90 cm by 60 cm. The basins benefit, particularly, poorer farmers with no access to draught power as they will not have delayed planting as they wait to borrow draught power from their neighbors (Mazvimavi and Twomlow 2007).

Mulch Ripping

Ox-drawn rip lines are made from attachments fitted on the plough frame and were developed to open furrows for moisture capture. They also break superficially compacted layers (Mandumbu 2011). Mapfumo et al. (2002) explained that mulch ripping makes use of the soil on the surface to protect the soil underneath, making soil disturbance limited to the planting zones, and ripping is done to a depth of 23 cm. In Zimbabwe, rip lines that are being promoted go to a depth of 23 cm.

Cotton Production Under Climate Change

Drought-tolerant cotton and in-field water harvesting are promising technologies to minimize the impacts of drought (Katengeza et al. 2019). The farmers in Zambezi valley in Zimbabwe have traditionally used two cotton varieties. Although these have over the years sufficed, they are currently failing due to recent trends on climate change. The low cotton genetic diversity and its use in traditional risk management may affect the resilience of farmers especially recently due to existing stresses which have made the environment more unpredictable (Meldrum et al. 2017). According to Thomas et al. (2015) the importance of crop genetic diversity or resilience and adaptation of farm systems to climate change is highlighted in many studies.

Recently, the cotton companies have promoted subsoiling and tied contours as in-field water harvesting techniques to cushion farmers against the adverse effects of climate change (Chaniwa et al. 2020). In situ water harvesting involves small movements of rainwater as surface runoff, in order to concentrate water where it is wanted most (Ibraimo and Munguambe 2007). Therefore the objectives of this study are to determine the effects of two tillage systems (conventional and tied ridges) on the performance of six cotton genotypes in semiarid northern Zimbabwe.

Effects of Water Harvesting on Soil Moisture Content

The results showed that tied rides had a significantly higher ($P < 0.05$) moisture content in the 0–5 and 6–10 cm depth compared to conventional system, while there were no significant differences between the two tillage systems at 11–15 cm (Fig. 4).

The results indicated that tied ridges had significantly higher percentage of moisture content at 0–5 and 6–10 but not at 11–15 cm. Higher moisture content at 0–5 cm illustrates the effectiveness of tied contours in moisture conservation on top soil horizons. The 2018–2019 season was a drought year in Zimbabwe, so the effects of the tillage method on moisture was apparent. These results are in tandem with Mupangwa et al. (2006) and Nyamadzawo et al. (2013) who reported the efficiency of tied ridges in capturing moisture and concentrating it on the root zone. Usually shallower depths are first ones to dry, hence the ability of the tied ridges to retain moisture makes the crop grow better compared to those in conventional plots; as the season was characterized by short duration, high-intensity rainfall, this might have

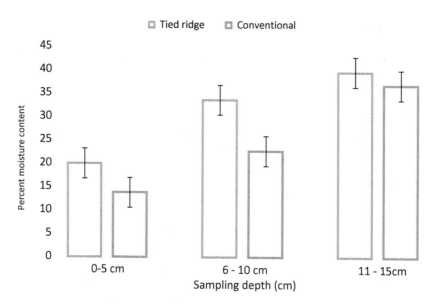

Fig. 4 Effects of tillage system on percentage moisture content at three sampling depths

resulted in moisture loss through runoff as rainfall intensity exceeded the infiltration capacity of the soil.

This means that the use of tied ridges as moisture conservation techniques was advantageous in the cotton as plants grown in tied ridges could still absorb moisture from the top horizons while those in conventional tillage could not. Higher moisture content will be followed by higher water use efficiency and subsequently higher yields. Promoting such a technology in the communities located in marginally drier areas would increase productivity of the cotton plants and help farmers cope with dryness caused by climate change.

Tied ridges increase the amount of water in the soil profile by trapping or holding the rainfall. These structures reduce runoff from the fields and enhance infiltration. This leads to higher amounts of stored moisture. Water harvesting is currently being rejuvenated in the marginal areas as an adaptive mechanism to climate change. High-intensity rainfall causes moisture to be lost through runoff. Research done by Nyamadzawo et al. (2012) found moisture losses as much as 50% being lost to runoff from cultivated fields. Due to the erratic nature of the rainfall such losses may never be recovered.

Soil water content near field capacity allows for best combinations of sufficient air space for oxygen diffusion, greatest amounts of nutrients in soluble forms, greatest cross-sectional areas for diffusion of ions and mass flow of water, and most favorable conditions for root extension.

The days to flowering varied significantly due to the effects of cotton genotypes. The results indicated that the genotypes M579 and M 577 had significantly fewer days to flowering compared to QM301 and CRIM 51 which are the Zimbabwe varieties (Fig. 5).

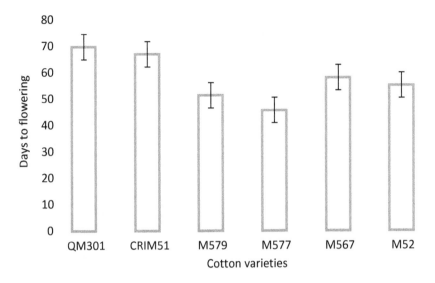

Fig. 5 The effect of cotton varieties on days to flowering

It was observed that M579 and M577 flowered the earliest compared to the rest of the varieties (Fig. 2). The variety M577 had the highest number of productive branches (Fig. 2). This according to Kooyers (2015) illustrates drought escape. Drought escape occurs when plants develop rapidly and reproduce when the environment becomes severe. Usually a variety that develops early escapes drought due to the shortened life cycle. The early flowering varieties represent a response when the plant's genetic resources are sourced to assist farmers to adapt to the changed climate. FAO (2015) reported that the biggest challenge for future food or commerce is to find a good match between the crops and the production environment as the effects of climate change increase. The two varieties are therefore suited to the changed conditions of semiarid Zimbabwe. These genotypes were obtained from India. Singh (2017) reported that one of the breeding strategies to mitigate the effect of climate change is to improve adaptation through increased access to a number of varieties at the local level with different growth durations to escape or to avoid predictable occurrences of stress at critical periods. This reduces the vulnerability of local farmers to the effects of climate change extremes.

The cotton variety M577 had the biggest cotton branch number compared to the rest of the varieties. The locally bred varieties showed lower branch numbers compared to the imported one (Fig. 6).

Branch number was also significantly ($P < 0.05$) affected by moisture conservation method. Tied ridges had significantly higher branch number compared to conventional tillage (Fig. 6).

The hybrids showed potential for use in environments with low water availability. Obtaining hybrids with good grain yield in environments with water restriction and a significant increase in water-limited environments has been an aim of many breeding

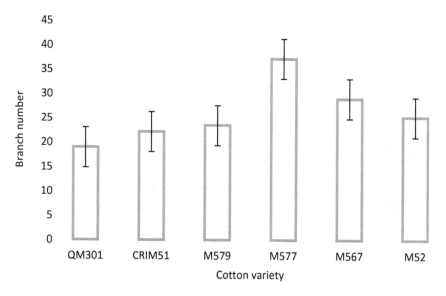

Fig. 6 The effect of cotton varieties on branch number

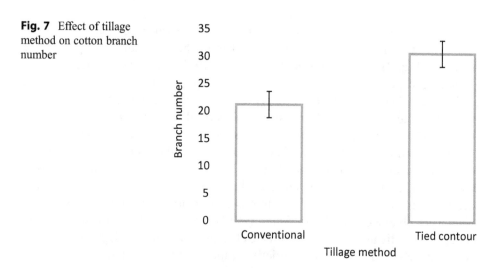

Fig. 7 Effect of tillage method on cotton branch number

programs. Varieties with lower yield reduction in stressful environment mean stability of production which is desirable.

A combination of high bearing branch number and few days to flowering is key in selecting crop varieties suited to climate change. One major characteristic of the seasons is erratic rainfall which begins early and terminates prematurely. Other seasons have delayed start and end prematurely. Generally, most seasons are not long enough to allow most varieties from germination until it reaches maturity. The varieties M577 and M567 suit such an environment as it can escape droughts through early phenological development. Quicker production of flowering branches also indicates tolerance to drought. Therefore, introduction of varieties to small-scale farmers in Zimbabwe is for farmers' adaptation to climate change.

For flower number the M577, M579, and M567 had the highest number of flowers. Flower number represents reproductive capacity. The same trend was observed on boll number. Higher flower numbers were observed on tied ridges compared to conventional system (Fig. 7). This showed that the varieties respond to tied ridges compared to conventional. As cotton is grown by small-scale farmers in high-risk environments, the results illustrate the suitability of these three varieties to the northern Zimbabwe environment. Adopting the three varieties to the Zimbabwean conditions illustrate Nhemachena et al. (2016)'s assertion that new crop varieties will be needed to cope with climate change. Areas with shortened seasons result in low yield potential, affecting agricultural productivity. Local breeding efforts may be slow to match the changing environments, hence the importation of genotypes that suit the environment. The new environment is forcing farmers to abandon some varieties or even abandon the crop especially if the varieties cannot withstand the new conditions.

The tied ridges gave a significantly higher branch number compared to conventional system (Fig. 5). This illustrates the suitability of the moisture conservation to maintain cotton production in arid areas which are vulnerable to climate change. Tied ridges tended to concentrate moisture on the few areas and that makes more moisture available for the crop.

The varieties M579 and M577 were the most suited to the climate of northern Zimbabwe as they performed much better that the traditional varieties which farmers have been using. Also making tied ridges improve water retention properties of the soil and leads to greater water use efficiency. Therefore adoption of tied ridges and the two varieties has potential to be an adaptive strategy that has potential to resuscitate an industry that was facing extinction due to climate change. Bringing new varieties and introducing water harvesting has been key in alleviating the effects of climate change on cotton production. The current conventional cotton varieties had failed to cope with climate change effects of reduced moisture availability. Other characteristics noted on the new varieties were very early development and increased branch number. These new varieties are better able to withstand the current adversities and keep marginalized farmers in production. The number of bolls noted per branch and on branches is all yield parameters which are critical for cotton productivity.

The cotton parameters noted above contribute to phenotypic plasticity which is the ability to cope with a wide range of environmental conditions. According to Kooyers (2015), drought escape may be optimal for annual plants in environments with shorter growing periods that are ended by sever terminal drought while drought avoidance may be optimal where if the growing season is punctuated by transient droughts.

References

Belay A, Gebrekidan H, Uloro Y (1998) Effect of tied ridges on grain yield response of Maize (*Zeamays L.*) to application of crop residue and residual N and P on two soil types at Alemaya, Ethiopia. S Afr J Plant Soil 15(4):123–129. https://doi.org/10.1080/02571862.1998.10635130

Brazier A (2015) Climate change in Zimbabwe: facts for planners and decision makers. Konrad-Adenauer-Stiftung, Harare

Buka G (2017) Cotton and its by-products sector in Zimbabwe. UNCTAD, Geneva

Chaniwa M, Nyawenze C, Mandumbu R, Mutsiveri G, Gadzirayi CT, Munyati VT, Rugare JT (2020). Ending poverty through affordable credit to small scale farmers: the case of Cotton Company of Zimbabwe. In: Nhamo et al. (eds) Scaling up SDGs implementation. Sustainable development goals series. https://doi.org/10.1007/978-3-33216-7_8

Christensen JH, Hewitson B, Busuioc A, Chen A, Gao X, Held I, Jones R, Kolli RK, Kwon W-T, Laprise R, Magaña Rueda V, Mearns L, Menéndez CG, Räisänen J, Rinke A, Sarr A, Whetton P (2007) Regional climate projections. In: Solomon S, Qin D, Manning M, Chen Z, Marquis M, Averyt KB, Tignor M, Miller HL (eds) Climate change 2007: the physical science basis. Contribution of working group I to the fourth assessment report of the intergovernmental panel on climate change. Cambridge University Press, Cambridge, UK/New York

FAO (2012) Climate change adaptation and mitigation. FAO, Rome

FAO (2015) Coping with climate change-the role of genetic resources for food and agriculture. FAO, Rome

FEWSNET (2019) Late and below normal rains and economic hardships expected to impact poor households food access. USAID

Fischer G, Shah M, van Velthuizen H (2002) Climate change and agricultural variability, a special report, on Climate change and agricultural vulnerability, contribution to the world summit on sustainable development. Johannesburg 2002 (Global, agriculture)

Global Agricultural Information Network (2017) Cotton production and consumption in Zimbabwe. USDA Foreign Agric Service

Ibraimo N and Munguambe P (2007) rainwater harvesting technologies for small scale rainfed agriculture in arid and semi-arid areas. Paper presented at the Integrated water resource management for improved livelihoods. CGIAR challenge program on water and food. South Africa

Katengeza SP, Holden ST, Lunduka RW (2019) Adoption of drought tolerant maize varieties under rainfall stress in Malawi. J Agric Econ 70:198–214

Kooyers NJ (2015) The evolution of drought escape and avoidance in natural herbaceous populations. Plant Sci 234:155–162

Mandumbu R (2011) Seedbank dynamics in conservation agriculture in south western Zimbabwe. Lambert Academic Publishers, Germany

Mapfumo P, Chuma E, Nyagumbo I, Mutambanengwe F (2002) Farmer field schools: facilitator manual; integrated soil water and nutrient management in semi-arid Zimbabwe. FAO of UN, Zimbabwe

Mazvimavi K, Twomlow SJ (2007) Conservation farming for agricultural relief and development in Zimbabwe. ICRISAT, Matopos Research Station, Bulawayo

McHugh OV, Steenhuis TS, Abebe B, Fernandes ECM (2007) Performance of in situ rainwater conservation tillage techniques on dry spell mitigation and erosion control in the drought-prone North Wello zone of the Ethiopian highlands. Soil Tillage Res 97:19–36

Meldrum G, Mijativic D, Rojas W, Flores J, Pinto M, Mamani G, Condori E, Hilaquith D, Grumberg H, Pdulosi S (2017) Climate change and crop diversity: farmers' perception and adaptation on the Bolivian Altipano. Environ Dev Sustain. https://doi.org/10.1007/s10668-016-9906-4

Mujeyi K (2013) Viability analysis of smallholder cotton production under contract farming in Zimbabwe. Invited paper presented at the 4th international conference of the African association of agricultural economists, September 22–25, 2013, Hammamet

Mupangwa W, Love D, Twomlow SJ (2006) Soil water conservation and rainwater harvesting strategies in semi-arid Mzingwane catchment Limpopo Basin, Zimbabwe. Phys Chem Chem. https://doi.org/10.1016/jpce.2006.08.042

Nhemachena C, Matchaya G, Nhlengetwa S, Nhemachena CR (2016) Economic aspects of genetic resources in addressing agricultural productivity in the context of climate change. In: Lal R et al (eds) Climate change and multidimensional sustainability in African agriculture. https://doi.org/10.1007/978-3-319-41238-2_9

Nyamadzawo G, Nyamugafata P, Wuta M, Nyamangara J, Chikowo R (2012) Infiltration and runoff losses under fallowing and conservation agriculture practices on contrasting soils, Zimbabwe. Water SA 38(2):233–240

Nyamadzawo G, Wuta M, Nyamangara J, Gumbo D (2013). Opportunities for optimisation of infield water harvesting to cope with changing climate in semi-arid smallholder farming areas of Zimbabwe. Springer Plus 2:100

Rockstro¨m J, Falkenmark M (2000) Semiarid crop production from a hydrological perspective: gap between potential and actual yields. Crit Rev Plant Sci 19:319–346

Singh RP (2017) Improving seed systems resiliency and local level through participatory approaches to adaptation to climate change. Adv Plants Agric Res 6:15–16

Thomas M, Thepot S, Galic N, Jovanne-Pin S, Remone C, Goldringer I (2015) Diversifying mechanisms in the on-farm evolution of crop mixtures. Mol Ecol 24:2937–2954

Twomlow SJ, Hove L (2007) Is conservation agriculture an option for vulnerable households in southern Africa. Paper presented at the conservation agriculture for sustainable land management to improve livelihoods for people in dry areas workshop, 7–9 May 2007, Damascus

UNEP (1997) Source book of alternative technologies for fresh water augmentation in Latin America and the Caribbean. International Environmental Technology Center, United Nations Environment Programme, Washington, DC

UNFCCC (2007) Climate change impacts, vulnerabilities and adaptation in developing countries. UNFCCC, Bonn

4

Impacts of Climate Change on the Hydro-Climatology and Performances of Bin El Ouidane Reservoir: Morocco, Africa

Abdellatif Ahbari, Laila Stour and Ali Agoumi

Contents

Abstract

In arid and humid contexts, dams' reservoirs play a crucial role in water regulation and flood control. Under the projected climate change (CC) effects, even a pre-optimized management approach (MA) of a reservoir needs to be assessed in this projected climate. This chapter aims to assess the impacts of CC on

A. Ahbari (✉) · L. Stour
Laboratory of Process Engineering and Environment, Faculty of Sciences and Techniques, Hassan II University of Casablanca, Mohammedia, Morocco

A. Agoumi
Laboratory of Civil Engineering, Hydraulic, Environment and Climate Change, Hassania School of Public Works, Casablanca, Morocco

the Hydroclimatic (HC) variables of the basin upstream the reservoir of Bin El Ouidane (Morocco), and the effects on the performances of its preoptimized MA. The applied Top-Down assessment procedure included CORDEX climate projections, hydrological, siltation, evaporation, and management models. Concerning the HC variables, the results obtained concord with those reported in the literature in terms of trend, but not always in terms of intensity of change. On the other hand, the projections expected a decrease in the performances of the reservoir, except for criterion allocations' standard deviation, calibrated during the optimization. Also, interesting conclusions have been found like: the change in precipitation dominant form, the accentuation of the pluvial hydrological regime, the advanced snow melting due to the temperature increase. This chapter presents a typical case study on how to use climate projections for reservoir MA adaptation, without being highly and negatively influenced by the climate model uncertainties.

Keywords

Reservoir · Impact · CORDEX · Management · Performance criteria · Bin El Ouidane

Introduction

Dams' reservoirs play a crucial role in the attenuation of the intra and interannual water resources heterogeneity. In fact, either in arid or humid context, these hydraulic infrastructures permit the flood control and assure the satisfaction of different water demand. In Morocco, the dam construction policy had helped in mitigating the impacts of intra-annual precipitations variability, drought periods, and massive flood events. Nonetheless, according to global and national climate projections reports (IPCC 2014; MEMEE 2016), the potential impacts of climate change (CC) can alter partially or totally this role of water regulation. Moreover, the adaptation to future CC effects should be considered in both design and exploitation phases of dams' reservoirs. On one hand, the adaptation in the design phase should focus on the change of the available water supply (Nassopoulos et al. 2012), the frequency and intensity of floods (Tofiq and Guven 2014), and the sediment yield volumes (Bussi et al. 2014). On the other hand, the adaptation during the exploitation phase is more oriented toward the regular update of the MA, and of its parameters (Zhang et al. 2017). Nonetheless, the update of MAs should not be limited to historical data, but the use of future climate projections resulting from global and regional climate models (RCM) is a path that requires further investigation (Brown et al. 2012). Several methodologies were attempted to adapt water planning to CC (Borison and Hamm 2008; Morgan et al. 2009). Nevertheless, managers still have a suspicious attitude toward climate projections (Means et al. 2010). Nowadays, it is clear that with all the problems related to observational data quality and accessibility, some models' aspects inefficiencies, uncertainties, diversified range of projected impacts, greenhouse emissions scenarios diversity, and the deduction of conclusions from the usage of climate projections is risked and limited (Brient 2020; Sellami et al. 2016).

Other than the use of climate projections to assess a preupdated MA, they can also be beneficial to highlight some negative aspects of the management to correct, or other positive points to enhance.

This chapter presents an assessment of the potential impacts of future CC projections on the Hydroclimatic (HC) variables on the basin upstream the dam reservoir of Bin El Ouidane, and the impacts on the future performances of the preoptimized Management Approach (MA) of this reservoir. In addition, the chapter aims to converge to conclusions that can be useful in the enhancement of the preoptimized approach. Furthermore, a comprehensive procedure that can be generalized to adapt reservoir MAs of other reservoirs worldwide is proposed.

Overview of the Adopted Framework of Analysis

Presentation of the Dam of Bin El Ouidane and the Impacts Assessment Framework

The hydraulic complex of Bin El Ouidane-Ait El Ouarda (named hereafter "BEO-AEO") is located in the high Atlas of Morocco, and composed of the major 1500 Mm^3 reservoir of Bin El Ouidane (BEO), and its 4 Mm^3 compensatory reservoir Ait El Ouarda (AEO). BEO-AEO was constructed in 1953 to satisfy the water demand of: (1) irrigation perimeters of Beni-Moussa and Tassaout-Aval; (2) drinking water of Afourer, Beni-Mellal, and neighboring villages; (3) production of hydroelectricity; and (4) flood control (Fig. 1).

The BEO-AEO reservoir complex is located at the outlet of an upstream basin of 6500 km^2. The analysis of the past hydro-climatologic variables evolution within the

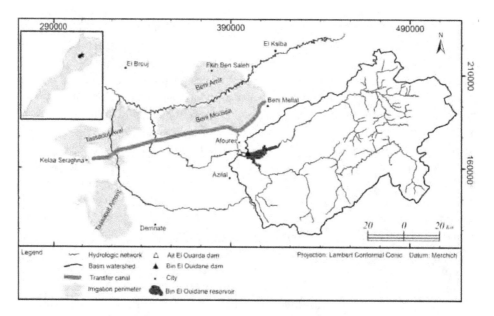

Fig. 1 Geographical map locating the basin of BEO, the dams, and the irrigated perimeters

basin of BEO (Ahbari et al. 2017) shows an augmentation trend of the annual temperature (+0,4 °C in 1989–2013 and of evaporation from the reservoir (+470 mm in 1976–2006), as stagnation of annual rainfall (around 400 mm during 1976–2014) and a substantial drop of the annual water supply to the reservoir by almost 50% after 1976 (from 1350 Mm^3 during 1939–1976 to 750 Mm^3 during 1976–2014).

In order to assess the impacts of future CC effects on the HC variables inside the basin located upstream the complex, the management variables of BEO-AEO and the performances of BEO-AEO, the typical Top-Down procedure described in Fig. 2 is used.

As shown in Fig. 2, the starting point is to choose a climate scenario. The appropriate way to proceed is by testing at least two opposite scenarios: one optimistic and another pessimistic. This will permit to have an idea about the variation interval of the eventual impacts. Then, the selected climate scenario will drive a climate model in order to output the future projections of temperature and precipitations for a given area. In general, the regional climate model is more suitable for the assessment of CC impacts assessment in basin-scale studies (Singh et al. 2019). After that, the resulting climate projections series will be used as direct entries for a prevalidated hydrologic and evaporation models. Once the projected water supply series are calculated, they will be inputted to a prevalidated siltation model to estimate the projected annual siltation volumes. The next step is to run the reservoir management model using the projected series from the hydrologic, siltation, and evaporation models, in association with the projected water demand. Finally, by using representative performance criteria, the results of the projected reservoir behavior are analyzed and compared to the performances outputted by the pre-optimized MA described in detail in Ahbari et al. (2019).

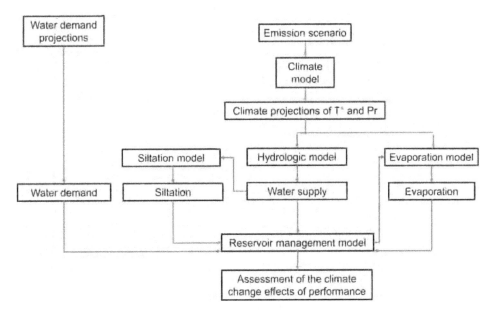

Fig. 2 Typical top-down procedure to assess the CC effects on reservoir performances

Since this chapter is more concerned about the daily management behavior, a daily time step is adopted. The reference period is between 1 March 1986 and 31 August 2015 (similar to the IPCC reference period for its AR5 report), and the projection period occupies the period between 1 March 2020 and 31 August 2049. The following subsections introduce the different components and models used in order to apply top-down assessment procedures to BEO-AEO.

Performance Criteria Used to Assess the Performances of the Reservoir

The Reliability

This criterion is calculated in relation to the duration of the demand satisfaction, as proposed by McMahon et al. (2006) in Eq. 1:

$$Rl = \frac{Nb\_of\_times\_D_t = 0}{n} \tag{1}$$

where D_t is the deficit at time step t; n is the number of time steps.

The Resilience

It is the probability that a state of demand's satisfaction follows a state of demand's dissatisfaction (Sandoval-Solis et al. 2011). The formula used is described in Eq. 2:

$$Rs = \frac{Nb\_of\_times\_D_{t+1} = 0\_follows\_D_t > 0}{Nb\_of\_times\_D_t > 0} \tag{2}$$

The Vulnerability

It is the likely deficit value of the system if a state of demand's dissatisfaction occurred (Hashimoto et al. 1982). In this chapter, the formula that estimates an average value of the likely deficit is adopted as shown in Eq. 3 (Sandoval-Solis et al. 2011):

$$Vul = \frac{\sum_{t=1}^{n} D_t / Nb\_of\_times\_D_t > 0}{water\_demand} \tag{3}$$

The Sustainability

It is a criterion that attempts to unify in a unique value the information driven by more than one performance criterion. The reformulated Looks (1997) sustainability formula suggested by Sandoval-Solis et al. (2011) is the one used in the chapter (Eq. 4):

$$Sus = \sqrt[3]{[Rl \times Rs \times (1 - Vul)]} \tag{4}$$

The Allocations' SD

This criterion permits to unveil to what extent a specific management approach has an unpredicted and unstable water allocations (Ahbari et al. 2019). Equation 5 details the calculation expression:

$$SD = \frac{\sqrt{\sum_{t=1}^{} (X_t - X_m)^2 / (n-1)}}{\text{Water\_demand}} \qquad (5)$$

where X_t is the allocated volume at time step t; X_m is the average allocated volume; n is the number of time steps.

Description of the Climate Models and Scenarios

As it was mentioned before, two representative concentration pathways (RCP) were used (Van Vuuren et al. 2011): the optimistic scenario "RCP 2.6" and the pessimistic scenario "RCP 8.5." The selection of the RCP 8.5 and 2.6 as assessment scenarios is driven by the fact that the objective of the chapter is to track the potential interval of plausible changes resulting from all levels of mitigation policies (no policy, weak, modest, below 2 °C policies). Therefore, the RCP 8.5 is chosen to represent the pessimistic trend, even though it is a highly unlikely case (Hausfather and Peters 2020), and the RCP 2.6 to simulate the optimistic tendency willing to keep the global warming below 2 °C. Concerning the climate projections, the data from the COordinated Regional Climate Downscaling EXperiment (CORDEX; Giorgi et al. 2009) were employed. The data output used are the bias-corrected climate projections over the EUR-11 domain which contains our studied area. The bias-corrected CORDEX data are accessible via the link https://esgf-data.dkrz.de/search/cordex-dkrz/. By respecting the same specifications during the data filtration (same Regional Circulation Model (RCM) and driven Global Circulation Model (GCM)) for both RCP scenarios, only two RCM were available for use: RCA4 and REMO2009. These RCM are driven respectively in their boundaries conditions by the GCM: ICHEC-EC-EARTH and MPI-M-MPI-ESM-LR. The NetCDF files containing the bias-corrected projections were downloaded for each variable-RCP-RCM triple, for the projection period 2020–2049. Since a global hydrologic model was used to simulate the streamflow in the basin, the calculation of the average projected series for the basin for each Variable-RCP-RCM triple was mandatory. So, the 6500 km^2 basin upstream BEO-AEO is divided into a 11 km grid cells, and the projected series of each triple variable-RCP-RCM at the center of each cell are extracted. The transformation of the cell center coordinates from regular to rotated was done using the software AgriMetSoft (2017). Then, the average basin series for each triple are calculated by averaging the projected series for all cells. To get the projected series for each variable-RCP couple, an ensemble mean was calculated using the series of the two RCM, and then converted to IS units (°C and mm).

Description of the Models Embedded in the Framework

The hydrologic model used is the hydrologic modeling system (HMS) model, with a simple canopy and surface formalisms, the soil moisture accounting (SMA) as the loss method, the Clark Unit Hydrograph (Clark UH) as the transform method, and the exponential recession method to simulate the baseflow component. The model was calibrated and validated over the basin in previous works (Ahbari et al. 2018a, b). The HMS model of the basin was obtained after a training–validation and a detailed sensitivity analysis processes that permit to converge to the best HMS model in terms of efficiency. Nonetheless, it is necessary to highlight that this rainfall–runoff model exhibits some deficiency in the simulation of some flow peaks as described in Ahbari et al. (2018b). This deficiency was taken into consideration while interpreting and discussing the results of this chapter. About the siltation process in BEO reservoir, it is modeled using a preestablished model (Ahbari et al. 2018c) integrating sediment yield to the reservoir and the sediment consolidation phenomena. As described in detail in Ahbari et al. (2018c), the preestablished siltation model was able to simulate the observed time series of siltation in BEO using a genetic algorithm optimization process. For the evaporation from the reservoir water plane, it is estimated using a multiple linear regression model between the monthly evaporated volume as a dependent variable and the monthly average temperature, the monthly potential evapotranspiration (ETP) and the reservoir water level at the beginning of the month, as explanatory variables. The mentioned evaporation model was calibrated and validated for the purpose of this assessment study, and the results are shown in the results section.

Concerning the BEO-AEO management model, it is represented by an optimized version of the current MA practiced in BEO-AEO. The optimization of the current MA was done using the genetic algorithm, and its performances were evaluated in a two-step process: a training-validation and a sensitivity analysis processes. For brevity purposes, more details about the reservoir operations fulfilled by this model please refer to Ahbari et al. (2019).

Evaluation of the Water Demand and the Impact Assessment Criteria

Due to the lack of detailed data about the projection of the daily water demand for each user for the projected period of 2020–2049, the series of the projected daily water demand was constructed by extrapolating the current hypothetical water demand, using the available projected evolution ratio of the basin of Oum-Er-Rabia. In fact, the basin of BEO-AEO is part of the basin of Oum-Er-Rabia for which the projected evolution ratios in 2020, 2025, and 2030 are available. Beyond 2030, the same ratio as for 2030 is applied till the end of the projected period. The current hypothetical water demand refers to an annual water demand series represented by the year on which the releases from BEO-AEO are maximal. This hypothetical series should be the best to represent the current water demand series (which is inaccessible for the authors) since the daily deficits will be at their minimal values ever recorded. Nevertheless, it is important to notice that the constructed annual water demand at the 2030 horizon (constructed using the concept described above) is almost equal to the only information publically available about the annual water demand for BEO-AEO for the same horizon (annual water demand at the 2030 horizon).

To evaluate the impacts of CC on the HC variables inside the basin of BEO-AEO and on its management variables, two assessment criteria were used: the average values between the reference and the projected periods, and the intra-annual evolution of the HC variables. Additionally, the following criteria were used to assess the change in the performances of the BEO-AEO complex in satisfying the water demand: the deficit, the reliability, the resilience, the vulnerability, the sustainability, and the standard deviation of allocations.

The Observed Impacts on Hydro-Climatology and Reservoir's Performances

The Elaborated Evaporation Model

The calibration period starts from March 1986 to February 2002, and the validation begins in March 2002 and finishes in December 2008. Figure 3 shows the linear regressions during the calibration and the validation periods.

As seen in Fig. 3, the linear regression is clear, important ($R^2 = 0.72$ for calibration and 0.76 for validation), and statistically significant at 5% level (p-value <0.0001). The resulting multiple linear regression relates the monthly evaporated volume from the reservoir of BEO to three explanatory variables as described in Eq. 6:

$$V_{EVP} = -46.03 + 0.06^*H_{WL} + 0.26^*T + 0.25^*ETP \qquad (6)$$

where V_{EVP} is the monthly evaporated volume (Mm^3); H_{WL} is the water level in the reservoir of BEO at the beginning of the month (m); T is the average monthly temperature (°C); ETP is the monthly potential evapotranspiration (mm). H_{WL} and T are both based on observed values of water level in the reservoir of BEO and the average temperature at the gauge of Tilouguite. The ETP was calculated based on the formulae (Eq. 7) dedicated for rainfall–runoff modeling purposes proposed by Oudin et al. (2005).

$$ETP = \frac{R_e}{\lambda.\rho}\frac{T_a + 5}{100} \Rightarrow \text{if } T_a + 5 > 0$$
$$ETP = 0 \Rightarrow \text{otherwise} \qquad (7)$$

where ETP is the rate of potential evapotranspiration (mm day^{-1}), R_e is extraterrestrial radiation which depends on latitude and Julian day (MJ m^{-2} day^{-1}), λ is the

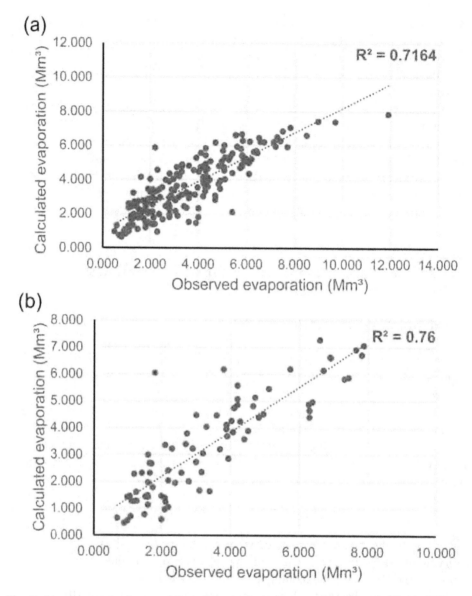

Fig. 3 Linear regression between observed and calculated evaporation during (**a**) calibration and (**b**) validation periods

latent heat flux equals 2,45 (MJ kg^{-1}), ρ is the density of water (kg m^{-3}), and T_a is mean daily air temperature (°C).

By integrating the water level as an explanatory variable, the obtained equation implicitly integrates the most important parameter that controls the evaporation from the reservoir: the surface of the water plane. The regression would be more substantial if other variables data were available like humidity and wind speed (Condie and Webster 1997).

CC Impacts on HC and Management Variables

Change in Average Values

Table 1 resumes the change in the average values of different HC and management variables between the reference and the projection periods.

The six key observations extracted from Table 1 are:

Firstly, negative impacts (qualified as negative because they alter directly or indirectly the performances of the reservoir) are suggested by both RCP for all variables, except for evaporation which will decrease. In fact, this first observation is an expected result and a concordance with the observed negative impacts of CC at both global (IPCC 2014) and national (MEMEE 2016) scales, especially for temperature and precipitation. Those negative impacts include: the increase of the average air temperature by at least +3.70 °C, the increase of ETP by at least 18.46%, the decrease of precipitation by 54.16% (RCP 2.6), the decrease of water supply by not less than 59%, the diminution of the stocked volume by 65.19% (RCP 8.5), and most importantly, the decline of water allocations to users to reach 60.07% (RCP 8.5). That said, what does not concord with global and national tendencies is the evaporation evolution trend. In fact, the downward trend of evaporation is explained by the decreasing trend projected for the variable water stocked in the reservoir, caused by the decrease of precipitations and water supply to the reservoir. The evaporation here is not to confound with the evaporation part of the evapotranspiration, which as it is seen will increase.

Secondly, except the temperature and the ETP which are related, the two RCPs drain practically the same impact on HC variables. Regarding this second observation,

Table 1 Change in the average values of the HC and the management variables between the reference and the projection periods for both RCP 2.6 and 8.5

		Reference period	Values		Difference (%) to the reference period	
		1986–2015	RCP 2.6	RCP 8.5	RCP 2.6	RCP 8.5
HC variables	Temperature (°C/day)	16.61	20.32	20.71	+3.70 °C	+4.10 °C
	ETP (mm/day)	2.96	3.51	3.56	+18.46	+20.08
	Evaporation (Mm³/day)	0.118	0.082	0.080	−30.52	−31.81
	Precipitations (mm/day)	1.16	0.53	0.53	−54.46	−54.16
	Water supply (Mm³/day)	2.25	0.90	0.88	−59.89	−60.82
Management variables	Stocked volume (Mm³/day)	599.89	221.02	208.84	−63.16	−65.19
	Allocated volume (Mm³/day)	1.88	0.797	0.75	−57.92	−60.07

the similar impacts of both RCP in terms of intensity and trend can be related to three reasons: (1) the horizon of projection 2020–2049 is probably far enough to sense a difference between an optimistic and pessimistic scenario in terms of warming, but it is not far enough to let the RCP 2.6 decreasing emissions to take effect on other aspects of climate much more complex like precipitations (Giorgi et al. 2019); (2) The used RCM and their driven GCM models are likely capable of simulating air temperature directly correlated to greenhouse gases concentration, but still have problem with the simulation of climate variables indirectly related to these gases like precipitations (Schliep et al. 2010); (3) since the projected period is located in the first half of the twenty-first century, it is expected that the difference between scenarios remains small (IPCC 2014; Hawkins and Sutton 2009). This last reason can be more highlighted by looking at the variables evaporation and water supply (obtained using the validated evaporation and hydrologic models), for which the difference between the RCPs is more important, even though they both integrate in their calculation the climate models outputs. For the ETP, more difference between the two RCPs is obtained since it was calculated by the formula of Oudin et al. (2005), which relates the ETP to the temperature directly without requiring any other climate variable. Moreover, the difference between the ETP values for the two RCPs is statistically insignificant at 5% level (p-value = 0.058), which supports the two reasons mentioned.

Thirdly, the difference between the optimistic and the pessimistic RCP is more pronounced for the management variables than the HC variables: Concerning the third observation, it highlights a valuable conclusion: if following the RCP 2.6 emissions limits will not have immediate and important repercussions on the climate system (IPCC 2014; Marzeion et al. 2018), it would be more likely beneficial to other systems like the capacity of dam's reservoir to regulate water volumes interannually. This conclusion can be supported by the more interesting difference between RCP 2.6 and RCP 8.5 impacts on stocked and allocated volumes. In addition, the cause behind this difference between RCPs impacts for management variables is that the optimized MA has acquired some adaptation capabilities during the sensitivity analysis done in Ahbari et al. (2019). In fact, the optimized approach was defined after testing different scenarios of pluviometry (low, moderate, and high) and water demand (normal and high scenarios).

Fourthly, unlike other variables, the RCP 8.5 would induce less impacts than the RCP 2.6 for the precipitation variable. With regard to this fourth observation, statistical significance tests were applied to the difference between the RCP 2.6 and 8.5 precipitation values, and found it statistically insignificant at 5% level (p-value = 0.878). The 95% confidence interval on the difference between the two RCPs values of precipitation is]−0.047; 0.040[. Therefore, one cannot be sure about the cause responsible of this apparent anomaly: Is it the RCM and the GCM used or it is just a question of sampling.

Fifthly, even though directly related, the precipitations and the water supply variables are not similarly influenced by the RCP selected. In other words, the fifth observation stipulates that while the precipitation is more impacted by the RCP 2.6, the water supply variable is more influenced by the RCP 8.5. This may sound intriguing, because since the two variables are highly related then they are supposed to behave similarly when they are trained by a given RCP scenario. However, this apparent anomaly could be explained by these two reasons: (1) as it was mentioned before, the difference between the RCP 2.6 and 8.5 for the variable precipitation is not statistically significant; (2) also, the difference between the two RCPs for the variable water supply is statistically insignificant. Thus, the causes behind this situation can be, in addition to a problem in RCM or GCM functionalities, a problem in statistical sampling related to the projection period chosen.

Sixthly, the percentage of change between the reference and the projection periods for all variables is very high. About this last observation, it is necessary to note that all variables values for the RCP 2.6 and 8.5 are statistically significant compared to the reference period at 5% level (p-value <0.0001). This means that according to the used RCMs, a statistically significant CC during the period 2020–2049 is expected in the basin upstream BEO-AEO compared to the reference period 1986–2015. Moreover, a statistically significant negative impact on the management variables will take place during the same period. For the high intensity of difference between reference and projection periods, it should be known that the values obtained include the uncertainties related to the climate, hydrological, siltation, and evaporation models. So, unlike the direction of change trend, the values of change obtained require a delicate interpretation before making any conclusion.

Intra-annual Evolution of the Variables

As demonstrated in the subsection before, the projections expect a substantial change in the average value of all studied HC variables. However, it is necessary and beneficial to analyze the intra-annual evolution of this change, to detect the most impacted months, and to deduct constructive conclusions for some eventual adaptation strategies. Figure 4 represents the intra-annual evolution of the average air temperature in the basin of BEO-AEO.

It is clear from Fig. 4 that the increase of temperature would not concern all months, and the difference between the two RCPs monthly temperature is practically zero. In fact, the hydrological year can be divided into three zones, differentiated by the intensity of temperature increase. Thus, three zones are recognizable: (1) zone 1 (December–January–February) characterized by a very limited increase; (2) zone 2 (November and March) which surrounds zone 1, and it shows a moderate increase of temperature; (3) zone 3 (April to October) known for its high-temperature increase compared to the other zones.

This zonation in the temperature augmentation was also observed in other basins using different RCM and GCM (Bannister et al. 2017; Santer et al. 2018). According to these case studies, the zones' length can change from one month to several months, but the common conclusion is that the winter season had never shown any increase of temperature. The reasons proposed to explain this unique behavior are: (1) climate models are deficient in simulating the Arctic oscillation and the north hemisphere winter climate (Cohen et al. 2012); (2) the high monthly bias was found

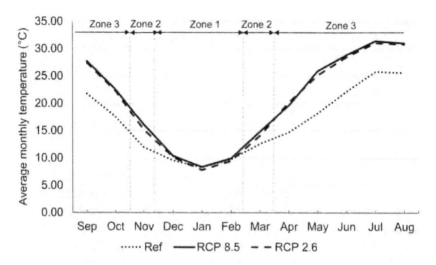

Fig. 4 Intra-annual evolution of the average air temperature under RCP 2.6 and 8.5 in the basin of BEO-AEO

correlated to climate models characterized by sparse horizontal resolution, and thus, to topographical factor (Bannister et al. 2017); (3) the surface-cloud feedback process might be incorrectly represented in climate models (Jiang et al. 2015); (4) the natural cause is not excluded either, because Osuch and Wawrzyniak (2016) have found a comparable seasonality in temperature evolution via the analysis of historical data.

Concerning the difference between the RCPs impacts on monthly air temperature, it was found that the difference between the RCP 8.5 and 2.6 temperature augmentation is statistically significant for November and May only (at 5% level, p-value equals 0.002 and 0.005, respectively). On the other hand, the monthly temperature increase compared to the reference period is statistically significant for all months, apart from January and February (p-value equals 0.818 and 0.626, respectively). These 2 months belonging to zone 1, confirm the characteristics of zone 1 showing, in terms of increasing values, a very limited change.

Like the air temperature, the ETP intra-annual variability presents the same zonation configuration (Fig. 5). Nonetheless, in the case of ETP, the zones' disposition has changed. Hence, the very limited increase zone (zone 1) covers December, January, and February (the change is statistically insignificant for January and February with a p-value reaching 0.926 and 0.672, consecutively at 5% level). Then, comes the moderate increase zone (zone 2), which gained some months previously occupied by zone 3 in Fig. 4 (March, April, September, October, and November). Finally, there is zone 3 starting on May and finishing on August, which shows a high increase of ETP compared to other months. All the differences between the reference period and the RCPs are statistically significant for all months covered by zones 2 and 3.

In reality, the ETP in this study was calculated using a formula that requires only the temperature and the latitude. By consequence, it was expected that the ETP intra-

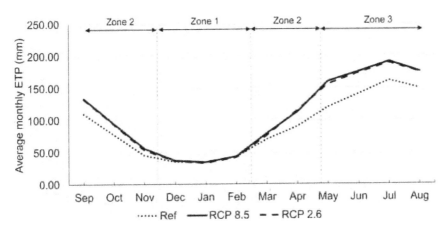

Fig. 5 Intra-annual evolution of ETP under RCP 2.6 and 8.5 forcing in the basin of BEO-AEO

annual variability would be similar to the temperature one. Hence, one can wonder why the pattern of intra-annual variability of temperature and ETP is different? In fact, this is related to the other parameter of the ETP calculation equation: Latitude. Thus, the translation of zones in Fig. 5 compared to Fig. 4 is related to the influence of the latitude parameter, and to its implicit factors such as the global radiation received by the studied area.

Other than the impact zonation, Fig. 5 indicates also that the differentiation between RCP 8.5 and 2.6 impacts on monthly ETP is visually hard to spot. In fact, statistical tests confirm this observation for all months, November, December, and May excepted (p-values equals 0.002, <0.0001 and 0.005 respectively). Hence, if an issue of climate modeling reason is expulsed, and the natural cause confirmed, this may indicate an interesting classification of months in terms of their ETP responses to greenhouse gases emissions. Consequently, once confirmed, this conclusion might be useful to take into consideration in the adaptation of reservoir operations approaches, so that they became prepared to future CC effects. The zonation remark is also to consider in those adaptation methodologies, since it is repeated in both temperature and ETP results.

For the variables temperature and ETP, the months covered by each zone can be seen differently by each reader, but it is sure that a zonation representing different intra-annual variability is present, and should be investigated further, and eventually taken into consideration in future adaptation strategies.

Regarding the intra-annual variability of evaporation from the reservoir of BEO, Fig. 6 describes a different monthly response of this variable compared to temperature and ETP. In fact, instead of showing months where the change is zero or very limited and others more impacted by CC, the evaporation intra-annual variability concerns all months with a slight difference in intensity from one month to another. That said, it is easy to spot months (December to March) where the difference between projections and reference curves is more pronounced than others (May to September). This is explained by the evolution of water stocked in the reservoir throughout time, it self-controlled by water supply and precipitations seasonality.

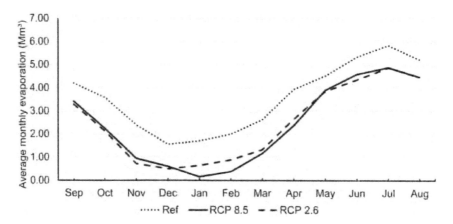

Fig. 6 Intra-annual evolution of evaporation under RCP 2.6 and 8.5 forcing in the reservoir of BEO

As seen, the downward trend of the average evaporation changes detailed in the subsection '3.2.1 Average values change' is detected in all months without exception, in opposition to the variables temperature and ETP. The monthly evaporation diminution shown in Fig. 6 are all statistically significant for all months at 5% level, with p-values ranging from 0.006 for September to <0.0001 for months between October and April. On the other hand, the difference between the two RCPs impacts on monthly evaporation seems to follow a differentiation by months. Hence, there is November, January, and February where the difference between RCP 2.6 and 8.5 is visible and statistically significant (p-value equals 0.044 for the first and <0.0001 for the lasts), and the other months where the type of scenario do not apply any statistically relevant difference.

With regard to the variable precipitations, Fig. 7 resumes the evolution of it under reference, RCP 2.6 and RCP 8.5 scenarios. By reading Fig. 7, four fundamental remarks can be deducted: (1) the diminution tendency affects all months, under the two RCPs; (2) the period covered by the rainy season is getting short, especially for the RCP 8.5; (3) The months with very low pluviometry are more frequent (For instance, from June to September); (4) the RCP 8.5 expects a delayed rainy season, while the RCP 2.6 suggests that it would be advanced temporally.

These remarks mean that the basin of BEO-AEO will experience a change in terms of the dominant precipitations type depending on the RCP considered (more solid for the 2.6 RCP, and more liquid for the 8.5 RCP), a change in the intensity and/ or frequency of summer storms (since the precipitations will be reduced considerably during this season) and of course all the repercussions that will be sensed in the hydrologic regime.

Statistically speaking, although the visual difference between the RCP 2.6 and 8.5 curves for some months (November and March for example) is clear, no significant difference was found for any month.

About the water supply to BEO-AEO and its variation through time, Fig. 8 compares the intra-annual evolution during the reference period and under the

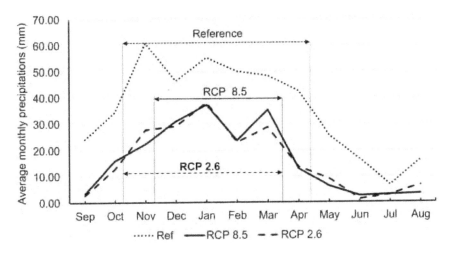

Fig. 7 Intra-annual evolution of precipitations under RCP 2.6 and 8.5 forcing in the basin of BEO-AEO

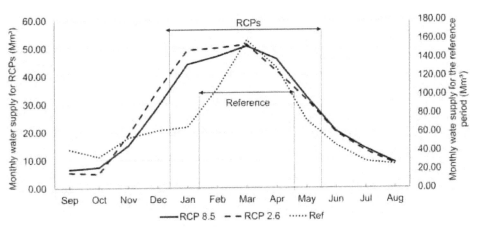

Fig. 8 Intra-annual evolution of water supply under RCP 2.6 and 8.5 in the basin of BEO-AEO

RCPs scenarios. The evolution of water supply concords with the conclusions of Fig. 7 and Table 1, assuming that future water supply will decrease in terms of volume and temporal distribution. In addition, Fig. 8 demonstrates that in parallel with the reduction of its volume drained, the season of "high supply" will prolong temporally compared to the reference period of 1986–2015. Moreover, the change of hydrologic regime from pluvio-nival to pluvial, mentioned in Ahbari et al. (2017), will be accentuated in both RCPs. Thus, the hydrologic peak is located in March, after being culminating in April during the period 1939–1976 (Ahbari et al. 2017). The accentuation of the pluvial regime can be justified when observing the RCPs curves, which present a peak roughly in the form of a plateau, especially for the RCP 2.6, where the plateau covers the months January–February–March.

In addition, the difference during March between the water supply curves of the reference and the projection periods' is statistically significant (p-value <0.0001 at 5% level), while it was insignificant during the same month for the variable precipitations. This may indicate that the impact of climate projections on March water supply was not induced via a change in March precipitations, but it is caused by another factor. In fact, the presumed vector of change is an advanced melting of the solid precipitations that occurred in previous months (especially for the RCP 2.6 where the dominant form of precipitations is the solid one), accelerated by the projected temperature increase. Then, if this is the case, the conclusions mentioned by other studies and confirmed in this chapter, about the incapacity of the RCM and GCM models to simulate the winter temperature increase, are reinforced again.

Additionally, in contradiction to the observation about dominant precipitation seen in Fig. 7, the water supply evolutions for the RCP 2.6 and 8.5 do not match the type of regimes expected.

Even though the solid precipitations are expected in RCP 2.6, the hydrologic regime is almost the same as for 8.5 on which liquid precipitations are more dominant. This is explained by the fact that the solid precipitations will not wait too long to melt, since the increase in temperature already mentioned provoke an advanced melting, and will not wait until the spring season. If this explanation is correct, this means that the reasons stated before about the stability of temperature in the winter season are correct, and the models are really encountering issue with the simulation of climate during this season.

CC Impacts on the Performances of BEO-AEO Reservoir

After running the assessment procedure detailed in Fig. 2 for the RCPs 2.6 and 8.5, the performances of the optimized MA of BEO-AEO under CC projections were assessed, via the calculation of the six most important criteria. Table 2 details the results obtained at a daily time-step scale.

According to Table 2, except the standard deviation of allocations, all aspects of performance will be deteriorated under the two emissions scenarios. Therefore, the

Table 2 Average daily values of the performance criteria of BEO-AEO calculated for the reference and projection periods

	Values (%)			Difference to the reference (%)	
	Reference	RCP2.6	RCP8.5	RCP 2.6	RCP 8.5
Deficit	26.68	67.01	68.70	+151.16	+157.48
Reliability	25.50	8.58	7.96	−66.33	−68.77
Resilience	11.16	4.58	4.18	−58.98	−62.50
Vulnerability	47.76	75.87	77	+58.88	+61.24
Sustainability	24.59	9.82	9.15	−60.04	−62.789
Standard deviation of allocations	38.73	19.36	17.77	−50.01	−54.10

percentage of deterioration can reach +157.48% (RCP 8.5) for the daily deficit, −68.77% (RCP 8.5) for the daily reliability, not less than −58.98% (RCP 2.6) in the case of the resilience and sustainability and at least +58.88% (RCP 2.6) of vulnerability deterioration. Moreover, it is proved by Table 2 that the vulnerability is the less impacted aspect, and the deficit the most damaged one.

Oppositely, the standard deviation of water allocations will be improved and its daily value will fluctuate between 17.77% (RCP 8.5) and 19.36% (RCP 2.6). The reason explaining this improvement is that the establishment of the optimized MA included a phase where the optimization of the releases to users has been performed. So, performing this update permitted to the MA to acquire an attitude to keep the variability of allocations minimal, under stressing climate circumstances.

Furthermore, and unlike the HC variables, the type of the scenario chosen appears to have an effect more considerable on the performances displayed by the BEO-AEO system. This confirms the conclusion mentioned before that: if the RCP 2.6 emissions guidelines are followed, it is certain that the climate will need decades to restore its regular conditions, but the systems impacted by this climate (the reservoir operations in this case) will rapidly start to show substantial enhancement.

By consequence, in the case of BEO-AEO complex, the aspects of performance that will manifest intensively this attitude are the daily deficit and the resilience. In fact, the reason behind the fast response of the deficit criterion is similar to the reason behind the improvement of allocations variability. Regarding the resilience, it is related to the fact that the choice of this optimized approach as the best one (Ahbari et al. 2019), was preceded by a sensitivity analysis towards different conditions (including varying pluviometry and water demand scenarios). Therefore, the different scenarios tested had likely helped in adapting the approach to extreme conditions, and by consequence, they improved its resilience capacities. Furthermore, the rapid response of the resilience criterion to the change of RCP can also be explained by the fact that, as shown in Ahbari et al. (2019), the resilience of BEO-AEO was the aspect of performance on which the manager of this reservoir complex have focused while calibrating the MA for the last time. This may indicate that if a regular update of the current MA was done, by including the resilience criterion in the objective function, the resilience may show similar results as those shown by the standard-deviation of allocations.

Overall, the results of the CC effects assessment present a clear tendency to have negative impacts on both the hydro-climatology of the basin of BEO-AEO and on the performances of the complex. In fact, the impacts projected concord with those reported in the literature (McSweeney et al. 2010; Ouraich 2014; MEMEE 2016; Filahi et al. 2017) in terms of trend, but not always concerning the intensity of change. For instance, McSweeney et al. (2010) and Ouraich (2014) stated a range of temperature variation across Morocco oscillating between +1.5 and +3.5 °C for 2060, and a maximum decrease of precipitations reaching 52%. Nevertheless, MEMEE (2016) mentioned that by 2065, air temperature in Morocco would increase from +0.5 °C to +2 °C, and precipitations change is projected to vary between +10% and −20%, depending on the region and the RCP scenario.

Nonetheless, it is necessary to mention that while the projected temperature and precipitations are averaged over all the catchment areas, the temperature, precipitation, and ETP of the reference period are gauge-based series. Thus, this may be explaining some of the differences. Additionally, water supplies are also based on average precipitation over the whole basin during the projection period, while in the reference it is based on the average of six gauges. Moreover, since all the performance values are based on water supply, then even these performance evolution compared to the reference period is to take with precautions.

The differences observed could be caused by the bias-corrected climate projections used in this study. In this chapter, the use of the CORDEX bias-corrected and downscaled data was dictated by the fact that neither the availability of meteorological stations inside and around the basin (especially for temperature), or the accessibility to those data was possible at the time of completion of this study. Therefore, the use of the available ones in the training and testing of a statistical downscaling and/or bias-correction method will never converge to acceptable results. These problems of network representativeness and limitation on climate model validation were reported by several authors (Filahi et al. 2017). So, the intensity of change in these conditions will never be accurate, even without counting the uncertainties driven by the different models and data.

Nevertheless, the differences can also be related to the various uncommon aspects between the materials and methods used in this chapter, and those employed in other studies, including: (1) the choice of RCM, GCM, and all the specifications that go in parallel (domain type, spatial resolution, ensemble calculation...) (Fantini et al. 2018; von Trentini et al. 2019); (2) the climate and hydrological models characteristics (Lespinas et al. 2014); (3) the downscaling and bias correction methods used and all the uncertainties they drive during training and testing (Teutschbein and Seibert 2012; Räty et al. 2018); (4) the choice of the reference, the projection periods, and the time step of projection (daily, monthly, or annually) used; (5) the area studied by each chapter (whole country, neighboring basin, neighboring city...).

In addition, to attenuate this disagreement between impacts assessment studies, it is recommended to proceed to the accomplishment of the following guidelines: (1) simple RCM assessment via a validation of the RCM simulations over a historical period; (2) advanced RCM assessment via a sensitivity analysis of RCM tuning parameters (Bucchignani et al. 2016); (3) The use of ensemble RCMs simulations instead of single RCM outputs, due to the benefits proven by this option (Phillips and Gleckler 2006; Filahi et al. 2017); (4) the assessment and selection of the best ensemble RCMs prediction method (Duan and Phillips 2010) without being influenced by the eventual negative impacts of ensemble mean RCM use.

However, this disagreement between studies, about the magnitude of change, is not only specific to the basin of BEO, but it is well-known worldwide at catchment scale (Sellami et al. 2016). Furthermore, the projections of CC can also differ in sign not only in magnitude (Koutsouris et al. 2010; Nassopoulos et al. 2012).

In addition, the intensity of change is not the ultimate objective in CC assessment studies, nor it is the aim of this chapter. But, the important is to prove to managers that impacts are coming, and adaptation actions are highly recommended. Moreover, the manager should know that even with an optimized MA, without regular updates, the future performances will be heavily affected. In fact, with all the uncertainties trained by the climate, hydrological, and other models used, the use of these projections, directly, to adapt the current MAs would be useless, or even counterproductive.

In reality, the climate projections can be used indirectly by exploiting common conclusions like those mentioned in this chapter (the trend of change, the zonation in the impact of CC on the variability of temperature and ETP, the change in precipitation dominant form, the change in hydrologic regime, which months are more affected, which performance aspects are more responsive to adaptability, which ones are more impacted by CC...).

Hence, the current MA can be modified by adding additional parameters, and/or optimized by testing various operations conditions to take into consideration those expected change. Once those modifications are applied to the current MA, the manager can update his approach regularly to implement each time the trend of change and its intensity. Furthermore, the manager can look for the best update pattern of this new approach, by comparing different update configuration (each year, each n years...).

For instance, in our example, before evaluating the impacts of CC on the performances of BEO-AEO, Ahbari et al. (2019) proceeded to the update and optimization of the releases formulae. Thus, when analyzing the results of performance projections, the standard deviation of water allocations was the only criterion which had experienced an improvement of its daily value.

Conclusions and Perspectives

The aim of the chapter was to assess the impacts of future CC projections on the HC variables, management variables, and the performances of the complex of BEO-AEO. To accomplish this objective, a typical top-down CC assessment procedure was followed including the use of optimistic and pessimistic emissions scenarios, RCM ensemble-mean, and prevalidated models (hydrological, siltation, evaporation, and reservoir management models). The climate projections data used were from the bias-corrected CORDEX data specific to EUR-11 spatial domain. Once the typical assessment procedure was run, the results were analyzed via: average values change, intra-annual variability, and performance criteria.

The results obtained show negative impacts for both HC (evaporation excepted) and management variables and all the performance aspects (the standard deviation of allocations excepted) of BEO-AEO. In general, the trends of HC variables change concord with other studies focusing on Morocco. Particularly, the projections expect an increase of air temperature and ETP, a downward tendency of precipitations, water supply, evaporation, stocked volume in the reservoir of BEO, allocated volumes to users. In terms of performances, and except the standard-deviation of allocations, negative trends of evolution are projected for all performance criteria: increase of deficit and vulnerability and decrease of reliability, sustainability, and resilience. Oppositely, the variability of water allocations will be improved under the two RCPs. This improvement is related to the phase of optimization of the releases that preceded this work. Concerning the intra-annual variability of the HC variables, the following conclusions are the most important ones: the zonation in terms of the impacts of CC on the variability of temperature and ETP, the change in precipitation dominant form depending on the RCP chosen, the accentuation of the pluvial hydrological regime, the probable advanced snow melting due to the increase of temperature, the limited difference between the impacts of the two RCPs on the majority of variables and months.

Finally, it is sure that the intensity of the projected CC mentioned in this chapter, and the one reported in other papers are not similar, but all of them have a consensus regarding the tendency of this change: Negative impacts are expected. Therefore, actions should be taken in order to adapt these infrastructures to CC effects.

References

AgriMetSoft (2017) Rotation of coordinates based on CORDEX domains [Computer software]. http://www.agrimetsoft.com/CordexCoordinateRotation.aspx. Accessed 10 May 2020

Ahbari A, Stour L, Agoumi A (2017) The dam reservoir of Bin El Ouidane (Azilal, Morocco) face to climate change. Int J Sci Eng Res 8(5):199–203. https://doi.org/10.14299/ijser.2017.05.013

Ahbari A, Stour L, Agoumi A et al (2018a) Estimation of initial values of the HMS model parameters: application to the basin of Bin El Ouidane (Azilal, Morocco). J Mater Environ Sci 9(1):305–317. https://doi.org/10.26872/jmes.2018.9.1.34

Ahbari A, Stour L, Agoumi A et al (2018b) Sensitivity of the HMS model to various modelling characteristics: case study of Bin El Ouidane basin (High Atlas of Morocco). Arab J Geosci 11:549. https://doi.org/10.1007/s12517-018-3911-x

Ahbari A, Stour L, Agoumi A et al (2018c) A simple and efficient approach to predict reservoir settling volume: case study of Bin El Ouidane reservoir (Morocco). Arab J Geosci 11:591. https://doi.org/10.1007/s12517-018-3959-7

Ahbari A, Stour L, Agoumi A (2019) Ability of the performance criteria to assess and compare reservoir management approaches. Water Resour Manag 33(4):1541–1555. https://doi.org/10.1007/s11269-019-2201-z

Bannister D, Herzog M, Graf HF et al (2017) An assessment of recent and future temperature change over the Sichuan Basin, China, using CMIP5 climate models. J Clim 30(17):6701–6722. https://doi.org/10.1175/JCLI-D-16-0536.1

Borison A, Hamm G (2008) Real options and urban water resource planning in Australia. Water Services Association of Australia occasional paper (20). Water Services Association of Australia, Melbourne

Brient F (2020) Reducing uncertainties in climate projections with emergent constraints: concepts, examples and prospects. Adv Atmos Sci 37:1–15. https://doi.org/10.1007/s00376-019-9140-8

Brown C, Ghile Y, Laverty M et al (2012) Decision scaling: linking bottom-up vulnerability analysis with climate projections in the water sector. Water Resour Res 48:W09537. https://doi.org/10.1029/2011WR011212

Bucchignani E, Cattaneo L, Panitz H-J et al (2016) Sensitivity analysis with the regional climate model COSMO-CLM over the CORDEX-MENA domain. Meteorog Atmos Phys 128(1):73–95. https://doi.org/10.1007/s00703-015-0403-3

Bussi G, Francés F, Horel E et al (2014) Modelling the impact of climate change on sediment yield in a highly erodible Mediterranean catchment. J Soils Sediments 14:1921–1937. https://doi.org/10.1007/s11368-014-0956-7

Cohen JL, Furtado JC, Barlow MA et al (2012) Arctic warming, increasing fall snow cover and widespread boreal winter cooling. Environ Res Lett 7:014007. https://doi.org/10.1088/1748-9326/7/1/014007

Condie SA, Webster IT (1997) The influence of wind stress, temperature, and humidity gradients on evaporation from reservoir. Water Resour Res 33(12):2813–2822

Duan Q, Phillips TJ (2010) Bayesian estimation of local signal and noise in multimodel simulations of climate change. J Geophys Res 115:D18123. https://doi.org/10.1029/2009JD013654

Fantini A, Raffaele F, Torma C et al (2018) Assessment of multiple daily precipitation statistics in ERA-interim driven med-CORDEX and EURO-CORDEX experiments against high resolution observations. Clim Dyn 51:877–900. https://doi.org/10.1007/s00382-016-3453-4

Filahi S, Tramblay Y, Mouhir L et al (2017) Projected changes in temperature and precipitation in Morocco from high-resolution regional climate models. Int J Climatol 37(14):4846–4863

Giorgi F, Jones C, Asrar G (2009) Addressing climate information needs at the regional level: the CORDEX framework. World Meteorol Organ Bull 58:175–183

Giorgi F, Raffaele F, Coppola E (2019) The response of precipitation characteristics to global warming from climate projections. Earth Syst Dynam 10:73–89. https://doi.org/10.5194/esd-10-73-2019

Hashimoto T, Stedinger JR, Loucks DP (1982) Reliability, resiliency, and vulnerability criteria for water resource system performance. Water Resour Res 18(1):14–20. https://doi.org/10.1029/WR018i001p00014

Hausfather Z, Peters GP (2020) Emissions – the 'business as usual' story is misleading. Nature 577:618–620

Hawkins E, Sutton R (2009) The potential to narrow uncertainty in regional climate predictions. Bull Am Meteorol Soc 90:1095–1107. https://doi.org/10.1175/2009BAMS2607.1

IPCC (2014) Working group I contribution to the IPCC fifth assessment report climate change (2014) the physical science basis. Cambridge University Press, Cambridge, UK/New York

Jiang D, Tian Z, Lang X (2015) Reliability of climate models for China through the IPCC third to fifth assessment reports. Int J Climatol 36(3):1114–1133. https://doi.org/10.1002/joc.4406

Koutsouris AJ, Destouni G, Jarsjö J et al (2010) Hydro-climatic trends and water resource management implications based on multi-scale data for the Lake Victoria region, Kenya. Environ Res Lett 5:034005

Lespinas F, Ludwig W, Heussner S (2014) Hydrological and climatic uncertainties associated with modeling the impact of climate change on water resources of small Mediterranean coastal rivers. J Hydrol 511:403–422. https://doi.org/10.1016/j.jhydrol.2014.01.033

Looks DP (1997) Quantifying trends in system sustainability. Hydrol Sci J 42(4):513–530. https://doi.org/10.1080/02626669709492051

Marzeion B, Kaser G, Maussion F et al (2018) Limited influence of climate change mitigation on short-term glacier mass loss. Nat Clim Change 8:305–308. https://doi.org/10.1038/s41558-018-0093-1

McMahon TA, Adeloye AJ, Sen-Lin Z (2006) Understanding performance measures of reservoirs. J Hydrol 324:359–382. https://doi.org/10.1016/j.jhydrol.2005.09.030

McSweeney C, New M, Lizcano G (2010) UNDP climate change country profiles: Morocco. http://country-profiles.geog.ox.ac.uk/. Accessed 18 Dec 2018

Means E, Laugier M, Daw J (2010) Decision support planning methods: incorporating climate change uncertainties into water planning. Water Utility Climate Alliance White Paper. https://www.wucaonline.org/assets/pdf/pubs-whitepaper-012110.pdf. Accessed 10 May 2020

MEMEE (2016) 3éme Communication Nationale du Maroc à la Convention Cadre des Nations Unies sur les Changements Climatiques. http://unfccc.int/resource/docs/natc/marnc3.pdf. Accessed 18 Dec 2018

Morgan M, Dowlatabadi H, Henrion et al (2009) U.S. Climate Change Science Program. Synthesis and assessment product 5.2. Best practice approaches for characterizing, communicating and incorporating scientific uncertainty in climate decision making. https://keith.seas.harvard.edu/files/tkg/files/sap_5.2_best_practice_approaches_for_characterizi.pdf. Accessed 10 May 2020

Nassopoulos H, Dumas P, Hallegatte S (2012) Adaptation to an uncertain climate change: cost benefit analysis and robust decision making for dam dimensioning. Clim Chang 114:497–508. https://doi.org/10.1007/s10584-012-0423-7

Osuch M, Wawrzyniak T (2016) Inter- and intra-annual changes in air temperature and precipitation in western Spitsbergen. Int J Climatol 37(7):3082–3097. https://doi.org/10.1002/joc.4901

Oudin L, Hervieu F, Michel C et al (2005) Which potential evapotranspiration input for a lumped rainfall–runoff model? Part 2 – towards a simple and efficient potential evapotranspiration model for rainfall–runoff modelling. J Hydrol 303(1–4):290–306. https://doi.org/10.1016/j.jhydrol.2004.08.026

Ouraich I (2014) Agriculture, climate change, and adaptation in Morocco: a computable general equilibrium analysis. PhD thesis, Purdue University

Phillips TJ, Gleckler PJ (2006) Evaluation of continental precipitation in 20th century climate simulations: the utility of multimodel statistics. Water Resour Res 42(3):W03202. https://doi.org/10.1029/2005WR004313

Räty O, Räisänen J, Bosshard T et al (2018) Intercomparison of univariate and joint bias correction methods in changing climate from a hydrological perspective. Climacteric 6:33. https://doi.org/10.3390/cli6020033

Sandoval-Solis S, McKinney DC, Loucks DP (2011) Sustainability index for water resources planning and management. J Water Resour Plan Manag 137(5):381–390. https://doi.org/10.1061/(ASCE)WR.1943-5452.0000134

Santer BD, Po-Chedley S, Zelinka MD et al (2018) Human influence on the seasonal cycle of tropospheric temperature. Science 361(6399):eaas8806. https://doi.org/10.1126/science.aas8806

Schliep EM, Cooley D, Sain SR et al (2010) A comparison study of extreme precipitation from six different regional climate models via spatial hierarchical modeling. Extremes 13:219–239

Sellami H, Benabdallah S, La Jeunesse I et al (2016) Quantifying hydrological responses of small Mediterranean catchments under climate change projections. Sci Total Environ 543(B):924–936. https://doi.org/10.1016/j.scitotenv.2015.07.006

Singh VL, Jain SK, Singh PK (2019) Inter-comparisons and applicability of CMIP5 GCMs, RCMs and statistically downscaled NEX-GDDP based precipitation in India. Sci Total Environ 697:134163. https://doi.org/10.1016/j.scitotenv.2019.134163

Teutschbein C, Seibert J (2012) Bias correction of regional climate model simulations for hydro-logical climate-change impact studies: review and evaluation of different methods. J Hydrol 12(29):456–457. https://doi.org/10.1016/j.jhydrol.2012.05.052

Tofiq FA, Guven A (2014) Prediction of design flood discharge by statistical downscaling and general circulation models. J Hydrol 517:1145–1153. https://doi.org/10.1016/j.jhydrol.2014.06.028

Van Vuuren DP, Edmonds J, Kainuma M et al (2011) The representative concentration pathways: an overview. Clim Chang 109:5–31. https://doi.org/10.1007/s10584-011-0148-z

von Trentini F, Leduc M, Ludwig R (2019) Assessing natural variability in RCM signals: comparison of a multi model EURO-CORDEX ensemble with a 50-member single model large ensemble. Clim Dyn 53:1963–1979. https://doi.org/10.1007/s00382-019-04755-8

Zhang W, Liuab P, Wangac H et al (2017) Reservoir adaptive operating rules based on both of historical streamflow and future projections. J Hydrol 553:691–707. https://doi.org/10.1016/j.jhydrol.2017.08.031

5

Barriers to the Adoption of Improved Cooking Stoves for Rural Resilience and Climate Change Adaptation and Mitigation in Kenya

Daniel M. Nzengya, Paul Maina Mwari and Chrocosiscus Njeru

Contents

Abstract

Majority of Kenya's citizens reside in the rural areas where wood fuel is still the primary source of energy for cooking. Continuing reliance on wood fuel against the backdrop of burgeoning population poses huge threats to the country's forest cover, undermining capacity for climate change mitigation and adaptation. This study

D. M. Nzengya (✉)
Department of Social Sciences, St Paul's University, Limuru, Kenya
e-mail: dmuasya@spu.ac.ke

P. Maina Mwari · C. Njeru
Faculty of Social Sciences, St Paul's University, Limuru, Kenya
e-mail: mainapaul72@gmail.com; chrocnjeru@gmail.com

conducted in Machakos and Laikipia counties explored; (i) women's perceptions of the health risks associated with dependence of firewood for cooking, (ii) women's attitudes and perceptions towards improved cooking charcoal stoves (ICS) as cleaner alternatives to traditional firewood stoves for cooking, and (iii) women's perceptions of barriers to adoption of improved cooking stoves. Study findings revealed that women were aware of the health risks associated with the use of firewood for cooking. However, despite these perceptions, upward trends in demands for firewood as a source of energy for cooking in the rural areas may persist in the next coming decades. Barriers to adoption of improved cooking stoves vary by sociocultural contexts. The study concludes that innovations that involve stakeholders especially participatory designs, monitoring, and evaluation of ICS might improve adoption levels. Moreover, innovations to increase adoption need to leverage on the opportunities provided by Sustainable development goal number 7 to accelerate adoption of ICS among other forms of cleaner, affordable, and sustainable sources of energy for cooking.

Keywords

Improved cooking stoves · Barriers to adoption · Rural resilience · Climate change mitigation · Women · Kenya

Introduction and Background

Sustainable development goal number seven, target 7.1 challenges the international development community and various governments to work towards achieving universal access to affordable, reliable, and modern energy services by 2030. It is estimated that about three billion people globally rely on biomass and coal burning for domestic use (WHO 2016). Majority of these population resides in developing nations (Clark et al. 2009). The solid biomass fuels available locally are such as wood, charcoal, dung, and agricultural residues for cooking, heating, and other domestic basic needs (Boampong and Phillips 2016). The emissions emanating from these energy sources pose health risks to humanity, among many other negative social and environmental impacts (Dickinson et al. 2015).

Ndegwa and others (2011) posit that 94% of the African rural population and 73% of urban population relay heavily on biomass energy especially for cooking and heating. They further state that wood fuel energy not only used as household energy but also used in schools, hospitals, colleges, small industries, and hotels. Women from rural areas and children below 5 years suffer most from health risks posed by biomass burning during long hours of cooking. The identified drivers of rising demands for wood fuel for cooking include, rapid population growth, inaccessibility to cheaper and affordable alternative energy substitutes, and the rising pervasive poverty and inequality.

Kenya has one of the largest rural populations that still rely on wood fuel as a primary source of energy for cooking. This preliminary study conducted Machakos and Laikipia counties explored; (i) women's perceptions of health risks and the socioenvironmental consequences associated with dependence of firewood for cooking, (ii) women's attitudes and perceptions towards improved cooking charcoal stoves as cleaner alternatives to traditional firewood stoves for cooking, and (iii) women's perceptions of barriers to adoption of improved cooking stoves by rural households.

Literature Review

Wood fuel is widely used in Kenya's rural areas and urban slums. It accounts for 68% of energy sources in the country with petroleum accounting for 22%, electricity at 9%, and coal at less than 1% (Ndegwa et al. 2011). It is estimated that biomass energy, firewood, charcoal, and agricultural wastes contributes approximately up to 70% of Kenya's energy demand and provides for almost 90% of rural household energy need. Dependence on wood fuel is among the contributors of deforestation exacerbating the country's greenhouse emissions. A study by Marigi (2017) has estimated that typical rural Kenyan family consumes about 11 kg of firewood daily, generating approximately 20.57 kg of carbon dioxide.

There is extensive research showing that traditional cooking stoves that utilize biomass contribute to a wide range of negative impacts on human health, air quality, and climate change. Although many agencies including World Health Organizations (WHO) recommend adoption of ICS in rural settings where cleaner sources of cooking are not available, to reduce indoor pollution, many communities are still challenged with appropriate stove selection, and sustained stove adoption and use. The efforts to better understand and provide solutions to this have faced several challenges. The challenges include matching new technologies with local socioeconomic conditions and cooking culture and designing comprehensive measurement strategies to effectively diagnose success or failure of these improved stoves.

Kenya is one of the countries where a great deal of research has continued to promote the adoption of improved cooking stoves in Kenya. Improved cookstoves have been recommended by many researchers as efficient and effective biomass-fuel stoves that reduce chronic obstructive pulmonary diseases, lung cancer, low birth weight, and premature mortality (Clark et al. 2009). In western Kenya, improved stoves were projected as human health survival especially to the children. The researchers promoted improved cookstove adoptions in the area with a goal to increase human survival through improving air, reducing disease, saving time and money, and reducing environmental degradation. Exposure to indoor air pollution has been blamed to poor human health especially women and both infants and small children who are present near the cooking site. Ezeh and others (2014) noted a 0.8% of neonatal deaths, 42.3% of post-neonatal deaths, and 36.3% of child deaths occurred in household using solid fuels for cooking; 70% of the deaths occurred in rural areas. Some of the promotional strategies in attempt to increase adoption of improved cooking stoves have included linking the prevalence of respiratory-related illness among household members to the indoor pollution associated with the use of firewood. These strategies have advocated for the use of improved cooking stoves as cleaner alternatives to mitigate indoor pollution, and research designs have revolved around generating empirical evidence to link adoption with improved health outcomes. Pilishvili et al. (2016) work showed that ICS were effective in reducing household air pollution leading to considerable acceptability in rural western Kenya. Also, the work by Silk and others (2012) work sought to increase adoption of locally produced ceramic cookstove in rural Kenyan household. Also, the work by Clark and others (2009) showed positive health impacts among women related to reduced indoor pollution following adoption of ICS.

Other researchers have attempted to leverage on ICS potential benefits of reducing indoor pollution to increase acceptability and adoption. Recent years have embraced much broader advocacy on the wider socioenvironmental benefits of ICS to entice communities' adoption. Jeuland and Pattanayak (2012) have investigated the benefits versus the costs of improved cook stoves in relation to wider implications of variability in health, forest, and climate impacts. The authors argued that adoption of ICS had broader benefits that included health improvement, time saving for households, preservation of forests, and reduction in emissions that contribute to global climate change. The authors concluded that households often find the improved cook stove technology to be inconvenient or culturally inappropriate resulting in disappointing uptake.

Liyama and others contend that ICS reduce deforestation and greenhouse gases that increase global warming leading to climate change. Other researchers who have explored related environmental benefits of ICS adoption include, Kiefer and Bussmann (2008) and Jeuland and Pattanayak (2012). These researchers have evaluated the benefits and costs of ICS on health, forests, and climate impacts, concluding that improved stoves provide improved health and time saving in household, preserve forests and related ecosystem and reduce emissions contributing to global climate change. On the contrary, biomass burning in inefficient cookstoves negatively impact on household level, community and national level, and regional and global level.

Both collecting and combustion of solid fuels used in household levels affects women and girls. Njenga et al. (2017) emphasize that scarcity of firewood forces women and girls to travel long distances looking for the commodity. The authors' claim has become part of the advocacy for improved cookstoves to ensure women could invest time in economically productive activities. Additionally, collecting and carrying is life threatening as women can suffer body injuries, rape, or attacked by wild animals (Njenga et al. 2017). Unsustainable firewood harvesting contributes to forest degradation, further contributing to loss in a nation's carbon sink. According to Loo et al. (2016), demand of biomass fuels is threatening forest cover given the rising demand for the commodity by schools, hotels, industries, among others, hence high emissions leading climate change.

The progress of achieving large-scale adoption and use of ICS has been slow and literature has little information of the slow uptake. Some researchers have contended that cost is not a barrier to adoption pointing that adoption and use of improved stoves is low even when households have given ICS without a charge. A study done in Peru, in 26 villages, only 46% of households used improved stoves given free of charge (Adrianzén 2010). The findings left the authors puzzled by the difficulties communities faced adopting ICS technologies given the benefits such as reducing household health burden related to indoor air pollution (Ezeh et al. 2014).

Materials and Methods

A questionnaire with open- and closed-ended questions was used to collect data from a sample of 54 women from the rural parts of Laikipia and Machakos County in 2018. In most rural areas in Kenya, cooking is one of the main chores for women. Data collected included energy sources for cooking for households, perceptions of health and environmental problems associated with the use traditional cooking stoves, extent of use of ICS, perceptions of socioenvironmental benefits, perceived importance to a household portfolio of the different sources energy for cooking, individual perceptions related to future trends of the usage of the different sources of energy for cooking, and perceived barriers to the adoption of ICS. Additional questions included sociodemographic information. Descriptive statistics were used to summarize data. Inferential statistics (independent t-test) were used to examine relationships between variables by region.

Results and Discussion

Sociodemographic Characteristics of the Sample

Table 1 below summarizes sociodemographic characteristics of the sample. Majority of the respondents in both counties were aged between 18–29 and 30–59 years. These are the productive age groups in a country and reliance on firewood collection and cooking takes away time that could be invested in productive economic engagements. The average household size for Machakos was 5.77, while that of Laikipia was 4.77. The differences in household size, number of primary school going children, and number of children under 5 years old between Machakos and Laikipia were statistically significant, t (52) $= 2.45$, $p < 0.05$, t (52) $= -4.74$, $p < 0.05$, and t (52) $= -6.06$, $p < 0.05$, respectively.

Source of Energy for Cooking and Perceived Importance of Various Sources by Study Sites

Majority of the respondents from the two study areas mentioned relying on both firewood and charcoal for cooking. The researchers further probed sources of energy for cooking. Majority cited that firewood was collected from nearby woodland resources (see Fig. 1). However, majority also mentioned buying both firewood and charcoal. During 1960s, majority of Kenya's rural population probably never imagined of a time when obtaining firewood would cost money, hence purchasing

Table 1 Sociodemographic characteristics of the sample

Variable	Category	Machakos % (Mean ± SD)	Laikipia % (Mean ± SD)
Age	18–29	15	25
	30–59	77	75
	60 or above	8	0
	18–29	15	25
Marital status	Married	77	57
	Single	12	21
	Separated	4	0
	Windowed	8	21
	Married	77	57
Average household size		5.77 ± 0.54	4.64 ± 1.72
Average number of school-going children		1.65 ± 1.9	3 ± 1.22
Average number of children under 5 years old		0.81 ± 0.83	2.46 ± 0.88

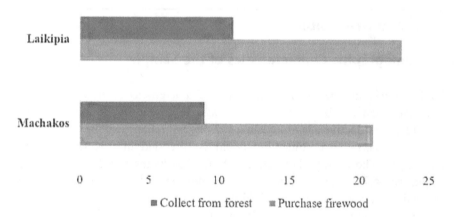

Fig. 1 Households' sources of firewood. (Source: Survey data)

firewood once regarded an abundant and free resource highlights that over the years sources of firewood have declined to the point that households must purchase the commodity. On average, respondents from Machakos spend 88.5 ($SD = 20.72$) per bundle of firewood, while those from Laikipia spend 237 ($SD = 50.00$) on firewood. On average, respondents from Machakos spend 48.40 ($SD = 4.95$) on charcoal, while those from Laikipia spend 57 ($SD = 5.39$). The differences in households' expenditures on both firewood and charcoal between the two counties were statistically significant, t (df) = 9.97, $p < 0.05$ (firewood), t (df) = 8.90, $p < 0.05$ (charcoal). Respondents were asked how much money they spend on a month to meet expenses related to energy for cooking. Results revealed that showed that on average, households in Machakos spend 2475 ($SD = 35.54$) compared to 875 ($SD = 144$) in Laikipia. The differences in monthly expenditures between the two counties were statistically significant, t (df) $= 6.67$, $p < 0.05$.

Continued reliance on firewood for cooking can be attributed to affordability and availability. In Laikipia, communities have access to adjacent forests for firewood. Moreover, the increase in vendors using motorcycles to ferry firewood further plays role in providing the commodity to households. Mobile vendors make it more convenient for households to save time that would otherwise be spent collecting firewood. It appears from the results obtained that firewood is a little more costly in Laikipia than in Machakos. This is probably because besides cooking, households in Laikipia also use firewood and charcoal for warming the house especially in the evening and during the cold season. It is also probable that most charcoal used in Laikipia is transported from distant sources, unlike in Machakos where residents rely on adjacent counties, Makueni and Kitui for the commodity.

Researchers asked respondents "who in the household was responsible for collecting firewood." Results obtained showed that women and children were involved in collecting firewood for cooking; however, men were largely involved in the collection and sale of firewood (see Fig. 2). Respondents were asked who in the household was responsible for paying for charcoal and firewood. Machakos, women households were responsible for meeting monthly energy expenses, while in Laikipia, men played that role. Few respondents mentioned children; these are probably elderly women who rely on their children or grandchildren for financial support.

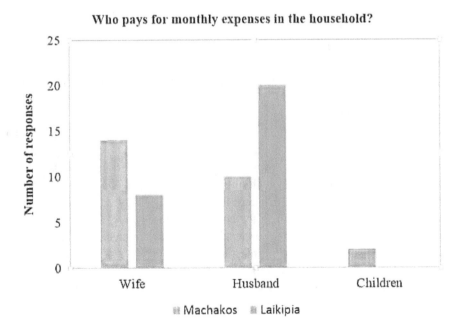

Fig. 2 Persons in the household responsible for paying household expenses related to energy used

In Laikipia, much of the responsibility of fetching firewood belongs to the wife; according to the findings of this study, it was observed that 78% of household expenses on firewood were done by men. Men in the region are involved more in forest destruction activities that include logging and charcoal burning. Sourcing of firewood and charcoal is expensive and all household budget for it. In Laikipia, majority of husbands play a significant role in meeting household needs mainly because they participate in preparing them (firewood and charcoal), transportation, and in some instances marketing. Another observation is that they buy firewood in larger quantities in form of sacks and large firewood bundles attracting higher value covered by majority of husbands. To some husbands, it is a full-time business.

In Machakos, it appears that the responsibility paying for fuel is perceived as feminine responsibility; majority of respondents at 53.8% admitted that it is the responsibility of the wife, 38.5% said it is husband's responsibility, while only 7.7% said it is the children who pays for fuel expenses.

Several researchers have noted the gendered division of labor when it comes to firewood collection and provisioning at the rural settings. Edelstein et al. (2008) have pointed on the gendered politics of firewood, contending that women have to transverse long distances to gather firewood. The implication is that women spend more time looking for energy sources instead of engaging in productive economic activities. Respondents were asked what other challenges beside increasing scarcity, women encountered when looking for firewood. One challenge that was frequently mentioned were increasing number of conflicts. All respondents from Laikipia reported experiencing conflicts associated with firewood collection. Conflicts over firewood may take a variety of forms, for instance, some sources are as a result of youth in the collection and sale of firewood for income. Other conflicts may be related to emerging rising demand from hotels, schools, small industries, and hospitals.

Table 2 summarizes respondents' perceptions in relation to the importance of various sources of energy for cooking in both Counties. The study observed that the main sources of energy for cooking is firewood and charcoal. This indicates rural population rely on trees for source of energy for cooking and warming, and therefore, there is need for energy saving technologies in both counties.

Results revealed that cow dung is not an important source of energy for cooking in Laikipia (90%) but an important a source of energy for cooking in Machakos (28%). Crop waste is not a source of energy in Laikipia (100%), but somehow a source in Machakos (42%). Due to high rate of population, Machakos may have faced acute deforestation situation, where in some places only cow dung and crop waste are the only source of energy for cooking and warming. Silk et al. (2012) pointed out that where fuel becomes expensive, the disadvantaged population turns to the available cheap ones including plastics.

Findings showed that also saw dust was somehow a source of energy in both counties: Machakos 38%, while Laikipia 17.8%. Kerosene is used for cooking in both counties: Machakos 80.7%, while Laikipia 100%. Kerosene is a common fuel even in rural small towns and can be purchased using <50 shillings. Handling kerosene stoves is easy despites its high production of particulate matter and carbon

Table 2 Perceived importance various energy sources for cooking by study site

Cooking energy source	Region	Not important (%)	Somewhat important (%)	Very important (%)
Cow dug	Machakos	72	12	16
	Laikipia	90	0	0
Crop waste	Machakos	58	31	12
	Laikipia	100	0	0
Saw dust	Machakos	58	39	4
	Laikipia	36	18	4
Firewood	Machakos	4	**35**	**66**
	Laikipia	0	**32**	**68**
Charcoal	Machakos	0	**39**	**62**
	Laikipia	0	**38**	**61**
Kerosene	Machakos	19	58	23
	Laikipia	0	46	54
LPG	Machakos	31	27	42
	Laikipia	0	89	11
Biogas	Machakos	50	46	4
	Laikipia	18	82	0

monoxide which endangers human health. A study by Pilishvili et al. (2016) observed that additional hour of kerosene use was associated with 5% increase in mean kitchen particulate matter. LPG is also a common source of energy in Machakos (69.2) and Laikipia (100%) as well as biogas recording 82.1% in Laikipia and 46.1% in Machakos.

Cow Dung and Crop Waste

In both regions, cow dung and crop waste according to the research indicate that majority saw it as not important. However, Machakos 16% indicated that it is a very important energy source. Machakos study indicates that some household of study have cleared most of the trees for other development activities. Therefore, the only available cheap fuel source is cow dung and crop waste which contributes high rate of indoor air pollution. This implies that higher health problems on both women and children in Machakos than in Laikipia. In Machakos, over 80% of the households has a cow and they live within the homestead; therefore, the energy source is always available but only preferred in time of extreme need. Promotion of improved cook stove in Machakos started many years back led by Nongovernmental organizations and the government. It therefore implies that majority of homes have an improved stove that easily used dried cow dung to cook. The climate in Machakos favors quick drying of the dung hence less time is needed to prepare it. The fact that 16% noted it as very important implies that the population entirely depends on cow dung for cooking. Use of crop waste is seasonal and fuel energy sources like maize, sorghum, and grass stalks are mainly available after harvest, therefore is an important energy source in Machakos. The pattern shows that those who regard it as very important are less than those who perceive it as somewhat important. It therefore implies that there is a certain population that entirely depends on crop waste for cooking and this indicates the possibilities of having serious biomass smoke-related health cases in Machakos than in Laikipia (Silk et al. 2012). In Laikipia, cow dung and crop residue are not important at all, since there are other better easily available energy choices. Also, the proximity to the forest enhances accessibility to the energy sources.

Kerosene, LPG, and Biogas

Kerosene as an energy source was noted as more important in Laikipia than in Machakos. This could be attributed to cost and availability since interviewed Laikipia residents seem more sound economically and price determines energy source to use (Silk et al. 2012). Again, one must have a stove designed to use kerosene. This therefore implies increased kerosene related indoor particulate matter pollution according to Pilishvili et al. (2016). In Laikipia, distribution networks for kerosene are more improved than Machakos, hence underlining the product mix concept that states that there is a correlation between products availability and consumption. Few Machakos women (23%) may be relying totally to Kerosene as a source of cooking, while for the rest, kerosene is mostly for light and biomass fuel for cooking. The study is from rural Machakos where few able household use electricity for light. Kerosene for cooking, compared to biomass fuel, is expensive for rural poor households. This could be the reason for low use of kerosene in Machakos.

LPG was noted as very important in Machakos than in Laikipia. However, those who stated as somehow important were at 89% meaning that they consider it necessary but it is not the only solution to energy needs. The price of a product like LPG determines the offering which the customers are willing to give to buy the targeted product. The importance attached to it in Machakos may imply that it is used for major cooking and different affordable packaging sizes in the market attracts many customers. In Laikipia, it seems the energy source LPG is used for light cooking mainly in the morning and late in the evening.

Biogas was noted as very important in Machakos as compared to Laikipia. However, on average, biogas was noted to be used more in Laikipia. This could be attributed by the fact that it requires relative huge investment to establishing a zero-grazing unit.

Women's Perceptions of Health Risks Associated with Reliance on Firewood

Participants were asked to list what they perceived as the health risks associated with the use of traditional firewood stoves for cooking. The lists of mentioned problems were entered in an Excel spreadsheet, and a tally was done to generate a graph summarizing frequently mentioned health risks/illnesses. Results obtained showed that the frequently mentioned health risks, in order of the most to least frequently mentioned health risks included: chest pain, sneezing, irritating eyes, breathing problems, and congestion of throat. Figure 3 summarizes the frequency of mention

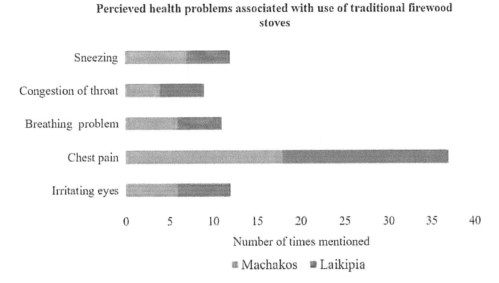

Fig. 3 Perceived health risks associated with the use of firewood traditional stoves for cooking

of the different health risks by respondent study site. Majority of women described chest pain to be a serious health risk. This was consistent across the two study sites. The secondly mostly mentioned health risk was sneezing, again patterns of the number of women describing sneezing problems were consistent across the two study sites. The third most mentioned health risk was eye irritation, and again, patterns of the number of women describing eye irritation problems were consistent across the two study sites. The least frequently mentioned health risk was congestion of throat, and again patterns of the number of women describing congestion of chest problems were consistent across the two study sites. Health risks mentioned can be summarized as respiratory and eye irritation illnesses. All these illnesses are associated with indoor pollution due to smoke and soot emanating from the burning of firewood. ICS are reported reduce indoor pollution related to smoke, and consequently, reduce the illnesses mentioned by the respondents in this study. Where adoption of ICS is challenge, some development agencies have innovated smoke hoods as options for reducing smoke, hence reduced levels of indoor pollution.

Perceived Future Trends on Use of Various Forms of Energy Sources

Respondents were asked what they perceived to be future trends on use of various forms of energy sources. Table 3 summarizes participants' responses by county. Results showed that almost all the respondents from Laikipia (100%) perceive the use of firewood to remain in the upwards trends (96.4%) compared to 46% from Machakos county. It is unclear what might inform this sort of perceptions. Respondents' perceptions correspond to the expected trends of firewood use in the coming decades, one would anticipate potential decline in negative impacts on forest cover further undermining climate mitigation measures and rural resilience (Njenga et al. 2017).

Table 3 Perceived future trends on use of various forms of energy sources by study site

Energy source	Region	Falling significantly (%)	Falling slightly (%)	Don't know (%)	Will remain about the same (%)	Rising slightly (%)	Rising significantly (%)
Cow dung	Machakos	54	9	15	4	19	–
	Laikipia	57	14	0	21	7	–
Crop waste	Machakos	46	12	8	19	8	8
	Laikipia	82	0	0	18	0	0
Firewood	Machakos	19	27	4	4	12	35
	Laikipia	0	0	0	0	43	58
Charcoal	Machakos	19	23	–	12	19	27
	Laikipia	0	0	–	4	14	82
Kerosene	Machakos	4	12	–	23	31	31
	Laikipia	0	64	–	32	4	0
LPG	Machakos	–	15	8	–	15	62
	Laikipia	–	0	0	–	61	39
Biogas	Machakos	12	19	2	4	19	23
	Laikipia	0	0	0	18	14	68

Participants from Laikipia, however, anticipate an upward trends in the use of LPG (100%). In addition, majority of respondents from Laikipia (82%) anticipate an upward trend in the adoption of biogas, compared to only 42.3% in Machakos county. LPG provides a capability to prepare meals in a shorter time, especially children's foods (Khadilkar 2015). Market for LPG have continued to provide attractive cylinder sizes affordable to low-income clientele. Also, majority of the women have an appeal for LPG as a clean energy for cooking, especially where mothers are accompanied by their children during cooking. Abundant crop waste, largely wheat and maize to support zero grazing is probably the reason for the perceived rise in biogas adoption in Laikipia.

Respondents from Machakos anticipate an upward trend in the use of kerosene. Besides cooking, kerosene also used for lighting in most rural homes in Machakos. Women overwhelmed with domestic chores prefer lighting a kerosene stove especially for lighter meals. Machakos has limited availability of crop residue and farm-based biomass. In Laikipia, respondents perceive a decline on the future use of kerosene for cooking. It is unclear what might inform this anticipation.

LPG use in Machakos is likely to rise significantly due to the ongoing government subsidy and use of flexible packaging containers. In Machakos, the current generation of population is constructing modern houses which have no kitchens to light firewood or charcoal giving preference to clean energies like LPG and biogas. In both counties, family sizes are becoming smaller with an average of two children, hence demand for heavy cooking will cease with time and demand for clean affordable energy like LPG and biogas will rise.

Perceived Barriers to Adoption of Improved Cooking Stoves

Respondents were asked on perceived barrier to adopt ICS. Table 4 summarizes responses by county. Results reveal that improved cook stoves are perceived to be costly compared to traditional stove. In both counties, Machakos (57.7%) and Laikipia (57.2) studies, cost was perceived to be a barrier to owning ICS. Some of these cook stoves are unaffordable by rural women especial those with no/little income. Jeuland and Pattanayak (2012) concur with this point when their study argued that women find it expensive to pay for the stove and self-learning of length of time to understand how properly to use the new technology.

Availability of charcoal affects adoption of ICS more in Laikipia (77.6%) than in Machakos (42.3%). The study shows that cost of charcoal affects ICS adoption in Laikipia (96.4%) more than Machakos (42.3%).

Stove design is probably not perceived as barrier in Machakos (34.4%); however, majority of the women from Laikipia (60.7%) seems to suggest design as problem. Stove designs acceptance depends on the provision benefits of cooking styles and needs of a given locality (Khadilkar 2015). Masera et al. (2007) cited the difficult of using some stove design to prepare traditional foods in Mexico. Also accepting a design means being able to use and repair. Therefore, improved cook stove may not be meeting the traditional needs of cooking in Laikipia. Also, low rate of literacy in

Table 4 Perceived barriers to adoption of improved cooking stoves by study site

Perceived hinderance	Region	Not a hinderance (%)	A little hinderance (%)	Don't have experience with this (%)	Somewhat a hinderance (%)	A very huge hinderance (%)
High cost	Machakos	12	19	12	15	42
	Laikipia	0	43	0	43	14
Availability of charcoal	Machakos	1	27	15	19	23
	Laikipia	18	4	0	68	11
Cost of charcoal	Machakos	15	27	15	15	27
	Laikipia	0	4	0	39	57
Design of stove	Machakos	35	27	8	27	4
	Laikipia	40	0	0	61	0
Availability of stove	Machakos	39	23	0	27	12
	Laikipia	0	0	0	32	68
Takes long to light	Machakos	27	31	8	15	19
	Laikipia	0	61	0	39	0
Takes too long to cook	Machakos	35	19	20	20	8
	Laikipia	0	43	57	0	0

the society may not be receptive of new technology ideas. Improved cook stoves availability is a barrier to own one in Laikipia county (100%), while in Machakos County (38.5%) is not an hinderance. Lighting ICS is not termed as a big issue in ICS ownership in both counties Machakos (30.8%) and Laikipia (60.1%). Cooking food using ICS in Machakos (34.6%) is not a barrier of owning one, but in Laikipia majority (57.1%) reported that they lacked experience of time taken in cooking. Sometimes new technologies are not accepted immediately because of the way users fear to change their behavior of cooking, other stove increase fuel consumption or increase time spend in cooking (Jeuland and Pattanayak 2012).

High Cost

In Machakos, high cost of stoves was noted as a major hindrance at 42% in stoves adoption. Today, due to challenges in livelihood income, every cent spent in a family counts and expenditure is based on priorities like paying school fees and buying food for the family. Possible reason could be that the stoves are never budgeted for and in some cases, it is perceived as luxurious and not a solution to indoor pollution. Most promoters of the stoves ask for prompt payment at once making them unaffordable. As mentioned earlier, domestic costs in Machakos are met by women with little or no support from men. Therefore, even if the stoves are sold at a discounted rate, they will always seem expensive. Also, the rural women may lack adequate information/awareness of benefits of improved cook stoves. ICS promoters in most cases mention short-term benefits such as saving firewood/charcoal but forget long-term benefits such health benefits. With this, concerned population end up missing crucial information especially health problems of children and mothers. The implication is that the adoption rate will be low, and firewood and biomass use continue creating negative impact to the environment.

Availability of Charcoal

This is also a major hindrance when it comes to stoves adoption in Machakos. Even though improved cook stove uses less charcoal, it must be of high quality and size to avoid smoking, and this is a challenge. In Machakos, charcoal is produced illegally, outside Charcoal Producer Association (CPA), hence inadequate in the market. For the last two decades, illegal charcoal production and land for agriculture degraded trees in rural Machakos and affected the availability of charcoal. The concept of demand and supply applies, if charcoal is not adequately available, it becomes expensive.

Cost of Charcoal

The reason for low adoption in Laikipia is noted as the cost of the charcoal at 57%. Possible reasons are due to the existence of other sources of energy which are cleaner and cheaper. Marketing of charcoal in Laikipia is mainly in bags and this to many is a huge cost.

Design of the Stove

In Machakos, the design of the stove was noted as a hindrance. Possible reasons could be the fact that stove design determines efficiency of cooking, size of cooking pot too, and the type of food to cook. Most of ICS are imported to Machakos rural areas and may be the designs do not meet the cooking needs of Machakos rural women. Again, some improved designs need specialist for repair which require funds when compared with three stone. Further, energy saving stoves are designed to save charcoal use, therefore their efficiency is lower and cannot be compared to the three stone firewood stoves. Stove design also influences the time taken to light the stove. According to the pattern from the research, stove design is a bigger hindrance in Laikipia. Availability of firewood and the need to warm houses during cold season are possible reasons. Improved cook stoves are made of vermiculite stone to reduce heat loss hence they do not warm the houses. The implication is that even if the stoves are acquired, they are never adequately used.

Availability of the Stove

This seems to be a huge hindrance in Laikipia, and the all the interviewees fell in the category of somewhat a hindrance (32%) and some stated it is a very huge hindrance (68%). Marketing of improved stoves is a huge business in the energy sector. Private companies have invested a lot and they need return on investment. Distribution of improved cook stoves is low in regions with adequate energy sources like charcoal and firewood, since few people buy them. This translates to low adoption rate and increased forest destruction to sustain the preferred energy sources.

Takes Too Long to Light

Respondents from Machakos (205%) seem to find the time it takes to light ICS to be problematic. Women engaged in household activities value time management. Due to heavy responsibilities within domestic arena, women are assisted to light stoves by their children. But children may face challenges lighting new designs. The comparison is based on traditional stoves. Lighting an improved stove requires a bit of skills that may be a challenge to the elderly. The problem is exacerbated by a lack of special gel for lighting cook stoves. To hasten lighting ICS, majority of the households use kerosene or old newspapers which adds to the cost/expenses of ICS for a household. The implication is that regular use of the stove does not happen.

Takes Too Long to Cook

Observed responses by study sample from Machakos county might be due to mostly preferred traditional diet in this region. Generally, improved cook stove is meant for light cooking. Quality of charcoal in Machakos is a factor since the best charcoal is from indigenous trees classified as hard wood which is rare in the region. Traditional foods are potentially most preferred due to economic challenges; however, cooking traditional dishes takes relatively longer compared to modern dishes. This has implications since ICS are relatively less used particularly among the elderly women.

Lessons Learned, Study Limitations, and Recommendations for Future Research

Even though access to cheaper sources of firewood continue to diminish, use of firewood for cooking still remains the most accessible form of energy for cooking in Kenya's rural areas. This trend is likely to persist in the next coming decade. Women are aware of the health risks associated with reliance on this source of energy. Given that low adoption of ICS as alternative methods of cooking may persist, there is need for interventions that reduce indoor pollution problems associated with continued reliance on firewood. This study was based on participants sampled from two counties; future research may need to include a representative mix of regional ethnic groups in Kenya. The variability on barriers to adoption highlight the need for a highly contextualized design and production of ICS to enhance adoption in culturally diverse contexts. Future research may need to explore the possibility of participatory design in which members of rural communities participate in designing ICS that appeal to their cultural preferences. This potential can be explored via the technical vocational institutions now almost distributed in every county in Kenya.

Conclusion

Citizens in Kenya's rural areas are aware of the health and negative socio-environmental consequences of relying on traditional unimproved firewood cooking stoves. However, there is still an inertia for continued dependence on firewood in the coming decades. Drivers of the perceived upward trends are a result of interaction of many factors. Barriers for adoption of ICS vary according to cultural taste and perceptions of ICS. Diminishing cheaper sources of firewood due to rapid population growth resulting to subsequent smaller land parcels imply future vulnerabilities by rural households as the cost of purchasing firewood is likely to keep rising, posing additional economic burden to households in terms of money spent purchasing firewood, but also on treatment of illnesses associated with indoor pollution from burning firewood. There is need to promote agroforestry as a potential option for meeting anticipated upward trends in demands for firewood to ensure resilience of rural livelihoods and sustainability of current efforts to improve the forest cover in Kenya.

References

Adrianzén M (2010) Improved stove adoption, firewood consumption and housewives' health: evidence from the Peruvian Andes. http://mitsloan.mit.edu/neudc/papers/paper_289.pdf. Accessed 12 Dec 2019

Boampong R, Phillips MA (2016) Renewable energy incentives in Kenya: feed-in-tariffs and rural expansion. University of Florida, Gainesville

Clark ML, Pell JL, Burch JB, Nelson TL, Robinson MM, Conway S, Bachand AM, Reynolds SJ (2009) Impact of improved cookstoves on indoor air pollution and adverse health effects among Honduran women. Environ Health Res 19(5):357–368

Dickinson KL, Kanyomse E, Piedrahita R, Coffey E, Rivera IJ, Adoctor J, … Hayden MH (2015) Research on Emissions, Air quality, Climate, and Cooking Technologies in Northern Ghana (REACCTING): study rationale and protocol. BMC Public Health 15(1):126

Edelstein MA, Pitchforth E, Asres G, Silverman M, Kulkarni NS (2008) Awareness of health effects of cooking smoke among women in the Gondar region of Ethiopia. BMC Int Health Hum Rights 8:1–7

Ezeh OK, Agho KE, Dibley MJ, Hall JJ, Page AN (2014) The effect of solid fuel use on childhood mortality in Nigeria: evidence from the 2013 crosssectional household survey.Environ Health. 13:113

Jeuland MA, Pattanayak SK (2012) Benefits and costs of improved cookstoves: assessing the implications of variability in health, forest and climate impacts. PLoS One 7(2):e30338

Khadilkar PD (2015) Capability approach-based evaluation of a biomass cook-stove design. Curr Sci 109(9):1601

Kiefer S, Bussmann RW (2008) Household energy demand and its challenges for forest management in the Kakamega area, western Kenya. Ethnobot Res Appl 6:363–371

Loo J, Hyseni L, Ouda R, Koske S, Nyagol R, Sadumah I, … Stanistreet D (2016) User perspectives of characteristics of improved cookstoves from a field evaluation in Western Kenya. Int J Environ Res Public Health 13(2):167

Marigi SN (2017) A stochastic investigation of solar energy application potential in rural households of Kenya and the associated environmental benefits. J Geosci Environ Prot 5(04):1

Masera O, Edwards R, Arnez CA, Berrueta V, Johnson M, Bracho LR, … Smith KR (2007) Impact of Patsari improved cookstoves on indoor air quality in Michoacán, Mexico. Energy Sustain Dev 11(2):45–56

Ndegwa G, Breuer T, Hamhaber J (2011) Woodfuels in Kenya and Rwanda: powering and driving the economy of the rural areas. Rural 45(2):26–30

Njenga M, Mahmoud Y, Mendum R, Iiyama M, Jamnadass R, De Nowina KR, Sundberg C (2017) Quality of charcoal produced using micro gasification and how the new cook stove works in rural Kenya. Environ Res Lett 12(9):095001

Pilishvili T, Loo JD, Schrag S, Stanistreet D, Christensen B, Yip F, … Bruce N (2016) Effectiveness of six improved cookstoves in reducing household air pollution and their acceptability in rural Western Kenya. PLoS One 11(11):e0165529

Silk BJ, Sadumah I, Patel MK, Were V, Person B, Harris J, … Quick RE (2012) A strategy to increase adoption of locally produced, ceramic cookstoves in rural Kenyan households. BMC Public Health 12(1):359

Osita Kingsley Ezeh, Kingsley Emwinyore Agho, Michael John Dibley, John Joseph Hall, Andrew Nicolas Page, (2014) The effect of solid fuel use on childhood mortality in Nigeria: evidence from the 2013 cross-sectional household survey. Environmental Health 13 (1)

6

Dual Pathway Model of Responses between Climate Change and Livestock Production

Adetunji Oroye Iyiola-Tunji, James Ijampy Adamu,
Paul Apagu John and Idris Muniru

Contents

A. O. Iyiola-Tunji (✉)
National Agricultural Extension and Research Liaison Services, Ahmadu Bello University, Zaria,
Nigeria
e-mail: tunjiyiola@naerls.gov.ng

J. I. Adamu
Nigerian Meteorological Agency, Abuja, Nigeria

P. A. John
Department of Animal Science, Ahmadu Bello University, Zaria, Nigeria

I. Muniru
Department of Biomedical Engineering, Faculty of Engineering and Technology, University of
Ilorin, Ilorin, Nigeria
e-mail: muniru.oi@unilorin.edu.ng

Abstract

This chapter was aimed at evaluating the responses of livestock to fluctuations in climate and the debilitating effect of livestock production on the environment. Survey of livestock stakeholders (farmers, researchers, marketers, and traders) was carried out in Sahel, Sudan, Northern Guinea Savannah, Southern Guinea Savannah, and Derived Savannah zones of Nigeria. In total, 362 respondents were interviewed between April and June 2020. The distribution of the respondents was 22 in Sahel, 57 in Sudan, 61 in Northern Guinea Savannah, 80 in Southern Guinea Savannah, and 106 in Derived Savannah. The respondents were purposely interviewed based on their engagement in livestock production, research or trading activities. Thirty-eight years' climate data from 1982 to 2019 were obtained from Nigerian Metrological Agency, Abuja. Ilela, Kiyawa, and Sabon Gari were chosen to represent Sahel, Sudan, and Northern Guinea Savannah zone of Nigeria, respectively. The data contained precipitation, relative humidity, and minimum and maximum temperature. The temperature humidity index (THI) was calculated using the formula: $THI = 0.8*T + RH*(T-14.4) + 46.4$, where T = ambient or dry-bulb temperature in °C and RH = relative humidity expressed as a proportion. Three Machine Learning model were built to predict the monthly minimum temperature, maximum temperature, and relative humidity respectively based on information from the previous 11 months. The methodology adopted is to treat each prediction task as a supervised learning problem. This involves transforming the time series data into a feature-target dataset using autoregressive (AR) technique. The major component of the activities of livestock that was known to cause injury to the environment as depicted in this chapter was the production of greenhouse gases. From the respondents in this chapter, some adaptive measures were stated as having controlling and mitigating effect at reducing the effect of activities of livestock on the climate and the environment. The environment and climate on the other side of the dual pathway is also known to induce stress on livestock. The concept of crop-livestock integration system is advocated in this chapter as beneficial to livestock and environment in the short and long run. Based on the predictive model developed for temperature and relative humidity in a sample location (Ilela) using Machine

Learning in this chapter, there is need for development of a web or standalone application that will be useable by Nigerian farmers, meteorological agencies, and extension organizations as climate fluctuation early warning system. Development of this predictive model needs to be expanded and made functional.

Keywords

Savannah · *Sudano-Sahel* · Climate change · Adaptation Livestock · Nigeria

Introduction

Livestock is important as sources of food (FAO 1993; Murphy and Allen 2003), fiber (Iyiola-Tunji 2012), and farm power (Srivastava 2006; Umar et al. 2013) in most part of sub-Saharan Africa. Adesogan et al. (2020) elaborated on the fact that the almost 800 million people who live in poverty (living on less than $1.90 per day) and subsist on a diet heavily based on starchy foods. They elaborated that animal source food will be required for millions more people who are slightly better off in terms of their incomes because animal source food provide not only calories but, more importantly, the nutrients required for achievement of human development potential. The dependability of some livestock keepers transcends the basic uses of the products and by-products of livestock to their uses as a form of savings for the raining days. Schmidt (2008) argued in favor of wealth storage in the form of cattle as a rational investment decision. Bettencourt et al. (2015) presented livestock feature as living savings which can be converted into cash whenever its needed, as well as a security asset influencing access to informal credits and loans and being also a source of collateral for loans.

It is expected that as the population of humans is increasing, the demands for animal products will also be increasing (FAO 2011). However, the production environments from which most of our animals are coming from in Africa are not improving commensurately to the potential demands for the stocks. The breeds of animals that are indigenous to specific locations in Africa have the advantages of adaptability to the environment from which they have lived for several hundreds of years. The environments to which these animals are adapted are heavily laden with stress. This in turn leads to low productivities. Heat stress is an intriguing factor that negatively influences livestock production and reproduction performances (Berihulay et al. 2019).

The dynamics of the environment in sub-Saharan Africa is widely varied within and between regions. In Nigeria, there are humid forest in the South and different categories of Savannah Northward. According to Abdulkadir et al. (2015), the potential impact of climate change, rainfall variability patterns and the dynamic hydrologic regimes have continued to escalate land degradation and make it imperative that the broad ecoclimatic zones could have changed. Variability of climate elements can also predispose animals to diseases. The distribution and incidence of animal diseases, specifically vector borne disease, are directly influenced by climate because the geographical distributions of vectors are predetermined by temperature and humidity (Kebede et al. 2018). Livestock production is being adversely affected by detrimental effects of extreme climatic conditions. Consequently, adaptation and mitigation of detrimental effects of extreme climates have played a major role in combating the climatic impact in livestock production (Khalifa 2003).

The level of aridity increases northward in the country. Haider (2019) reported on the challenges associated with climate change in Nigeria which are not the same across the country. The low precipitation in the North and high precipitation in parts of the South were reported to have led to aridity, drought, and desertification in the North and erosion due to flooding in the South (Onah et al. 2016; Akande et al. 2017). The more arid zones are the regions with the most population of livestock like cattle, sheep, and goats. Animals like camel and donkeys are exclusively found in the most arid regions of the country also. Over the years there had been reported cases of extreme high temperatures, drought, flooding, and some other climate-induced stressors. These phenomena always result in losses in productivity of the animals and accruable incomes to the farmers. So, in combating these problems, farmers (especially pastoralists) had adopted migration southward with their animals during the dry season when feed resources and water are not readily available. Some more adaptive measures along with the seasonal migration of stocks were evaluated in this chapter.

Apart from the effect of climate change on livestock which had been studied extensively, animals on higher production levels tend to be more sensitive to high temperature and humidity (Hahn 1989; Aydinalp and Cresser 2008; Nwosu and Ogbu 2011), there is also need for the understanding of the effect of livestock production activities that are capable of causing changes in climatic elements. Based on the submission of Brown (2019) and FAO report (http://www.fao.org/news/story/en/item/197623/icode/), rearing livestock generates 14.5% of global greenhouse gas emissions that are very bad for the environment. Livestock and their by-products account for million tons of carbon dioxide per year (Flachowsky and Kamphues 2012). Extensive system of livestock production plays a critical role in land degradation, climate change, water, and biodiversity loss. The problems surrounding livestock production cannot be considered in isolation. Economic, social, health, and environmental perspectives will be critical to solving some of these problems. There is need for development of a greater understanding of these complex issues so that we may encourage policies and practices to reduce the adverse effects of livestock production on climate, while ensuring that humans are fed and natural resources are preserved. A Human Society International report advocated that mitigating the animal agriculture sector's significant yet underappreciated role in climate change is vital for the health and sustainability of the planet, the environment, and its human and non-human inhabitants. Reducing greenhouse gasses (GHG) emissions, especially from animal agriculture is both urgent and critical (https://www.humanesociety.org/sites/default/files/docs/hsus-report-agriculture-global-warming-and-climate-change.pdf). This chapter however was aimed at evaluating the observed effects of fluctuation of climatic elements on livestock production and vice versa.

Climate Projection

The climate of the future is not clear due to how factors such as socioeconomics, technology, land use, and emissions of greenhouse gases will change and unfold (van Vuuren et al. 2011). A climate change scenario represents a specific possible future climate with for example high amounts of green technology contra a scenario with low amount of green technology. The dominant climate change scenarios are the representative concentration pathways (RCP) family of climate change scenarios. There exist mainly four RCP scenarios which are the RCP2.6, 4.5, 6, and 8.5. The two latter numbers indicate the radiative forcing target level for the year 2100 given a specific timeline, where the radiative forcing is the net change in the energy balance of the earth system due to some forcing agent expressed in watt per square meters (W/m^2) (Myhre et al. 2013; van Vuuren et al. 2011). These radiative forcers can be anthropogenic or natural, which can be greenhouse gas emissions or volcanic eruptions, respectively (Myhre et al. 2013).

The RCP2.6 trajectory signifies immediate anthropogenic intervention with strong climate change mitigation (van Vuuren et al. 2011). The RCP4.5 trajectory signifies stabilization of greenhouse gas emissions which like the RCP2.6 is also a scenario containing anthropogenic climate change mitigation but as prolific (Thomson et al. 2011). The RCP6 trajectory is similar to RCP4.5 but where climate change mitigation policies and technology implementations are not as strong (van Vuuren et al. 2011). The RCP8.5 trajectory signifies what is called as the "business as usual" trajectory with an increase in population, slow socioeconomic development, and slow innovation/implementation of technology (Riahi et al. 2011).

A core concept in the discussions around climate change is that of "adaptive capacity" or the potential of a society to adapt with the changes (if any) that might occur in the social ecological system from climate change (IPCC 2007a, b; McClanahan et al. 2008). Changes in climate have the potential to affect the agricultural industry which in turn can affect economic investment and population movements in countries. The livelihoods of many people, notably the poor and vulnerable, could be threatened if government and resource managers are not prepared for even the modest changes associated with climate change (Downing et al. 1997).

Climate of Nigeria

The climate of Nigeria is dominated by the influence of three main wind currents: the Tropical Maritime (TM) air mass, the Tropical Continental (TC) air mass, and the Equatorial Easterlies (EE) (Ojo 1977). The TM and TC air masses meet along the Inter-Tropical Discontinuity (ITD), which is a key driver of Nigeria's climate. The position of the ITD and oscillation during the year affects the spatial and temporal distribution of key climate characteristics of the country (Adegoke and Lamptey 1999). Following the annual movement of the ITD across the Equator, the rainfall season over Nigeria advances from the coast to the inland areas from March to

August and retreats from September to November, with a pronounced dry period between December and February. The rainfall patterns in Nigeria show the southern parts of the country with annual rainfall over 3000 mm and semiarid conditions in the north with annual rainfall less than 500 mm.

Materials and Methods

Survey of Livestock Stakeholders

A survey of livestock stakeholders (farmers, researchers, marketers and traders) was carried out in Sahel, Sudan, Northern Guinea Savannah, Southern Guinea Savannah, and Derived Savannah zones of Nigeria (Fig. 1). The regions under these ecoclimatic zones cut across all the 19 States and the Federal Capital Territory (FCT) of Nigeria. In total, 362 respondents were interviewed between April and June 2020. The survey instrument used was designed as an online questionnaire (for literate respondents). The other respondents who cannot fill the online form were administered printed questionnaire for the survey.

The distribution of the respondents was 22 in Sahel, 57 in Sudan, 61 in Northern Guinea Savannah, 80 in Southern Guinea Savannah, and 106 in Derived Savannah (Table 2). The respondents were purposively interviewed based on their engagement in livestock production, research or trading activities. The researchers were sourced

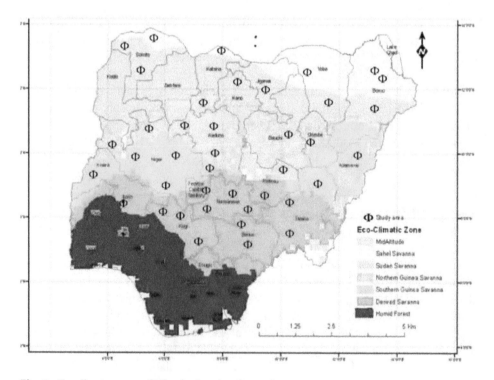

Fig. 1 Ecoclimate zones of Nigeria showing the study areas

Table 2 Production characteristics of livestock stakeholders in Savannah and Sudano-Sahelian zones of Nigeria ($N = 326$)

	Frequency	Percent (%)
Primary occupation of stakeholders		
Livestock farmers	203	62.3
Research scientist	92	28.2
Livestock marketer or trader	31	9.5
Type of animals being reared by respondents		
Cattle	99	30.4
Sheep	117	35.9
Goat	125	38.3
Donkey	5	1.5
Camel	12	3.7
Micro-livestock	72	22.1
Poultry	144	44.2
Preferred management system as indicated by respondents		
Intensive	163	50.0
Semi-intensive	133	40.8
Extensive	30	9.2
Distribution of respondents according to climate zones		
Sahel	22	6.7
Sudan	57	17.5
Northern Guinea Savannah	61	18.7
Southern Guinea Savannah	80	24.5
Derived Savannah	106	32.5
Awareness of the concept of climate change		
Yes	300	92.0
No	13	4.0
Maybe	13	4.0

N is the number of respondents

through their institutional affiliations. The farmers and marketers were sourced through the Agricultural Development Programs and Ministry of Agriculture (or Livestock) of the 19 States in the Northern regions of Nigeria as well as FCT. Key informant interview was conducted with Alhaji Ibrahim Mohammed – Director, FADAMA and Infrastructural Development of Yobe State Agricultural Development Program, Yobe State. The primary data obtained from this work were analyzed using frequency counts and percentages through crosstab analysis of Statistical Package for Social Sciences (SPSS) Version 16.

Climate Data and Analysis

Representative locations were chosen for Sahel, Sudan, and Northern Guinea Savannah. Ilela, Kiyawa, and Sabon Gari were chosen to represent Sahel, Sudan, and northern Guinea Savannah zone of Nigeria, respectively. Thirty-eight years' climate data from 1982 to 2019 were obtained from Nigerian Metrological Agency, Abuja. The data contained precipitation, relative humidity, and minimum and maximum temperature. This chapter employed the use of grid data obtained from the US National Oceanic and Atmospheric Authority (NOAA) reanalyzed historic data and complimented with Soil and Water Assessment Tool (SWAT) data. The major climatic parameters used in this chapter are rainfall, relative humidity, and temperature. To understand the nature of rainfall variation and trend and to determine climate extremes, data from 1982 to 2019 (38 years) for all weather stations within the study area were used. Descriptive statistical methods such as mean and standard deviation were utilized. Furthermore, time series was used for the analysis of rainfall trend over time, and the Moving Average Technique was also used in the analyses of the data. This chapter employed the use of the 3-Year Moving Average. The moving average has the characteristics of reducing the amount of variation in a set of data. This property in the time series is used mostly to remove fluctuations that are not needed. The use of moving average resulted in the formation of new series in which each of the actual value of the original series is replaced by the mean of itself and some of the values immediately preceding it and directly following it Ayoade (2008). To estimate the value of a variable Y (i.e., rainfall), corresponding to a given value of a variable X (i.e., time), regression analysis was applied. This was accomplished by estimating the value of Y from a least-squares curve that fits the sample data.

Standardized Precipitation Index and Trend Analysis

The Standardized Precipitation Index (SPI) calculation used was based on the long-term precipitation record for the desired period. This long-term record is fitted to a probability distribution, which is then transformed into a normal distribution so that the mean SPI for the location and desired period is zero (Edwards and McKee 1997). Positive SPI values indicate greater than median precipitation, and negative values indicate less than median precipitation. Because the SPI is normalized, wetter and drier climates can be represented in the same way, and wet periods can also be monitored using the SPI.

A correlation was done to determine how well a linear equation describes or explains the relationship between variables. From this analysis, the coefficient of determination was obtained, this is given by R^2. The standardized precipitation values were calculated for all the years from the use of the long-term mean, yearly mean, and the standard deviation using the equation below:

$$\varphi = \frac{X - \overline{X}}{\sigma}$$

where φ represents the standardized departure, x is the actual value of the parameter (annual rainfall), ẍ is the long term mean value of parameter (30 years rainfall average), and σ is the standard deviation.

Confidence test was performed on the dataset used and it was verified using 95% confidence interval. Coefficients of skewness, kurtosis, and variation were also investigated.

Temperature Humidity Index

The temperature humidity index (THI) was calculated using the following formula:

$$THI = 0.8 * T + RH * (T\text{-}14.4) + 46.4$$

where T = ambient or dry-bulb temperature in °C and RH=relative humidity expressed as a proportion, that is, 75% humidity is expressed as 0.75.

Results and Discussion

Rainfall Trend/Patterns in Nigeria from 1982 to 2019

The analysis shows the standardized rainfall anomaly over different climatic zones in Nigeria from 1982 to 2019. In the coastal, tropical rainforest, guinea, Sudan savannah areas, it was observed that there are more wet years than dry years. But for the Sahel savannah, the dry years were more than the wet years during the 48 years study period. The result corresponds to IPCC projection stating that the coastal areas are prone to more wet years leading to the occurrence of flooding and rainfall induced erosion, while region around the Sahel will experience more of drought as a result of reduction in the total precipitation.

Comparison of Variations in Climatic Elements Among Sahel, Sudan, and Northern Guinea Savannah Zones

Precipitation

Figure 2 showed the weighted average precipitation for Sahel, Sudan, and Northern Guinea Savannah zones of Nigeria. Ilela in Sokoto State was used as a reference point for Sahel while Kiyawa, Jigawa State and Sabon Gari, Kaduna State were used as reference points for Sudan and Northern Guinea Savannah zones, respectively. The number of months with substantial period of precipitation was seven from April to October at Ilela (Sahel). The maximum precipitation (7.86 mm) was recorded in July. Similar trends of duration of precipitation were also observed at Kiyawa and Sabon Gari. However, the maximum amount of precipitation was 11.97 and 12.31 mm, respectively, for Kiyawa and Sabon Gari. Figure 3 showed the average total precipitation (mm) for Sahel, Sudan, and Northern Guinea Savannah zones of

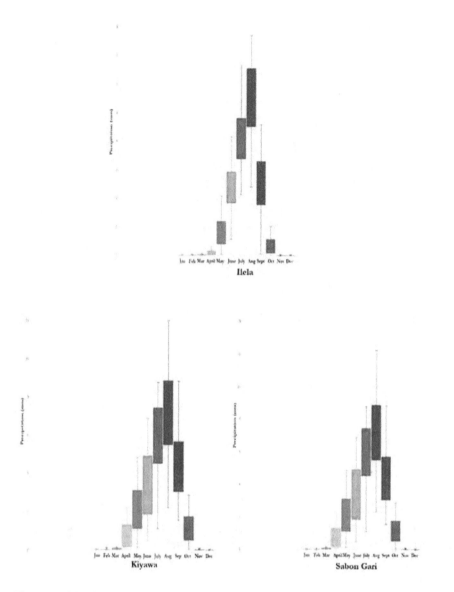

Fig. 2 Weighted average precipitation for Sahel, Sudan, and Northern Guinea Savannah zones of Nigeria for 1982 to 2019

Nigeria from 1982 to 2019. The average total volume of precipitations within the 38 years for the three zones were 614.79, 937.32, and 958.58 mm, respectively, for the Sahel, Sudan, and Northern Guinea Savannah zones. The volume of precipitation for Sudan and Northern Guinea Savannah were almost similar for most periods of the year except for July, August, and September when the volumes of rainfall was higher in Northern Guinea Savannah zones of Nigeria. The onset and end of rainfall in the two regions were similar.

Analyses of Standardized Precipitation Index (SPI) over the Sahel Savannah of Nigeria are presented in Fig. 5. The figure showed that in the first decade (1971–1980) and the second decade (1981–1990) the whole region had mostly

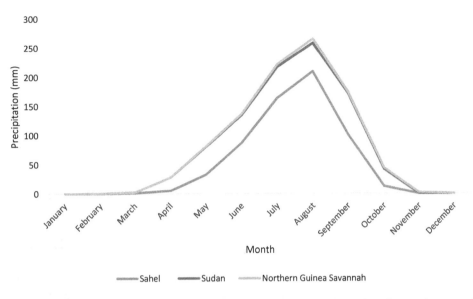

Fig. 3 Average total precipitation (mm) for Sahel, Sudan, and Northern Guinea Savannah zones of Nigeria for 1982 to 2019

negative anomalies. This indicates the zone suffered from serious hydrological drought from 1971 to 1990. However, there was a recovery to positive anomalies in the third decade (1991–2000), fourth decade (2001–2010), and the current decade (2011–2018). The dry years were more than the wet years during the 48 years study period. The result shows the region recorded 27 dry years and 19 wet years which corresponds to IPCC (2007a) projection stating that the Sahel will experience more of drought as a result of reduction in the total precipitation. With the predominant dry years in the region, water erosion should not have been a problem. Areas affected by water erosion challenges in the region indicates the little rainfall amount recorded occurred at very short interval with high intensity thereby generating runoff. This rainfall pattern is typical under a changing climate.

The analysis shows rainfall trend over Sahel Savannah of Nigeria for 1982–2019 as shown in Fig. 1. From 1981 to 1997 rainfall was increasing and decreasing in cycle of 4–5 years, though the cycle was in a declining rainfall order. During the first decade (1982–1990), the pattern showed decreasing rainfall amount. The second decade (1991–2000) up to 2018 showed a steady increase in rainfall amount a little above the average for region. This trend showed by the moving average for the region is in line with the work of Nicholson and Palao (1993), who reported that rainfall in West Africa generally decreased with latitude with essentially zonal isohyets.

Rainfall Trend/Patterns in Guinea Savannah of Nigeria

Analyses of Standardized Precipitation Index (SPI) over the Guinea Savannah of Nigeria clearly show that the first decade (1982–1991) had positive anomalies, and in the second decade (1981–1990) the whole region had mostly negative anomalies.

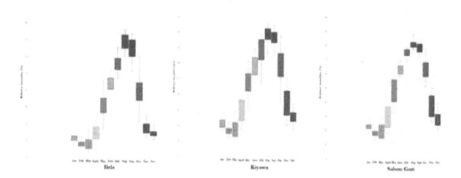

Fig. 4 Weighted average of relative humidity for Sahel, Sudan, and Northern Guinea Savannah zones of Nigeria for 1982 to 2019

However, there was a recovery to positive anomalies in 1991–2000, 2001–2010, and 2011–2018. The dry years were more than the wet years during the 38 years study period. The result shows the region recorded 22 dry years and 15 wet years which corresponds to IPCC projection stating that the region will experience more of wetness as a result of increase in the total precipitation. This is an indication of increased rainfall pattern in the Guinea Savannah region of Nigeria.

Figure 3 shows the rainfall trend over Guinea Savannah of Nigeria for 1971–2018. In the first decade (1971–1980) and the second decade (1981–1990), it was observed that rainfall was below normal (1971–2000) in the region. During the third (1991–2000), fourth (2001–2010), and current decade (2011–2018) it shows a steady increase in rainfall amount in the region above normal. This result is in line with the work of Nicholson and Palao (1993), who reported that rainfall in West Africa generally decreased with latitude with essentially zonal isohyets.

Relative Humidity

The variations of relative humidity for the zones being considered in this chapter are depicted in Fig. 4. The highest proportions of relative humidity were record in August in the three zones being considered in this chapter. However, the amount of water in the atmosphere was lowest in March of every year across the three regions as shown in Fig. 4. The highest values for relative humidity were 81.11%, 85.55%, and 88.06% in Sahel, Sudan, and Northern Guinea savannah zones, respectively. The lowest value also follows similar trend of decreasing northward the zones with 7.77%, 10.09%, and 12.53%, respectively, for Sahel, Sudan, and Northern Guinea savannah zones.

Atmospheric Temperature

Figures 5 and 6 show the minimum and maximum temperature in the Sahel, Sudan, and Northern Guinea Savannah zones of Nigeria as represented by Ilela, Kiyawa, and Sabon Gari. The highest value for minimum temperature was observed in May

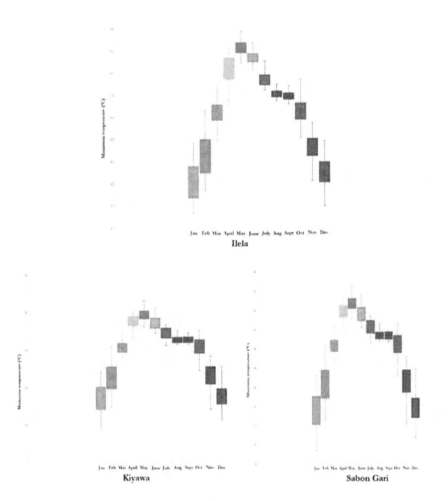

Fig. 5: Weighted average of minimum temperature for Sahel, Sudan, and Northern Guinea Savannah zones of Nigeria for 1982 to 2019

and the coldest temperature was in January. While the coldest temperature at Ilela, the Sahel climate, is 12.29 °C; the other two climate zones had similar values of 9.34 °C. Ilela had the highest value for the minimum temperature which was 28.76 °C which was followed by 26.55 °C recorded for Sabon Gari and the least among the three climate zones was 25.19 °C recorded by Kiyawa. The hottest average temperature recorded in all the three zones for the period under consideration in this chapter was 42.55 °C which was recorded at Ilela in the Sahel climate.

The Concept of Temperature Humidity Index

Figures 7 and 8 show the temperature humidity index as calculated using minimum and maximum temperatures, respectively. Animals, especially cattle, start having mild stress from index of 72 to 78. Severe stress starts from 79 to 88 (Table 1). Using the minimum temperatures as reference, the animals in Sahel ecoclimate were mildly

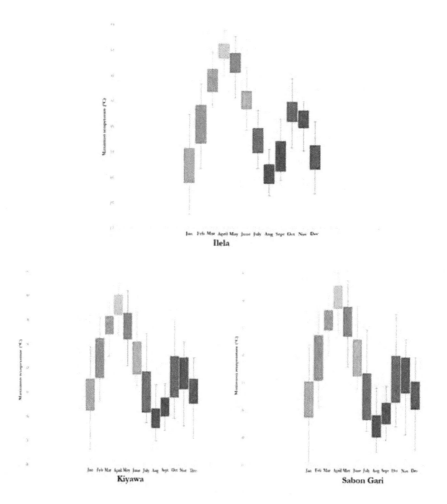

Fig. 6 Weighted average of maximum temperature for Sahel, Sudan, and Northern Guinea Savannah zones of Nigeria for 1982 to 2019

stressed due to heat and relative humidity interactions in May, June, and some days of July. Considering the animals during the maximum temperatures; they were mildly stressed in January, February, and December. That is the period of *harmattan* in the region. However, the animals are severely stressed for most of the other periods of the year. There were occasions of very severe stress on the animals during some parts of May and June (Fig. 8).

Livestock Production Characteristics in Sahel, Sudan, and Guinea Savannah Zones of Nigeria

Table 2 shows the production characteristics of livestock stakeholders in the Savannah and *Sudano-Sahelian* zones of Nigeria. Majority of the respondents were livestock farmers (62.3%). Substantial proportions of the respondents were research

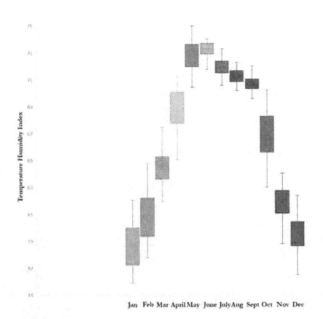

Fig. 7 Temperature humidity index using minimum temperature at Ilela, Sokoto State, as reference point for Sahel ecoclimate zone

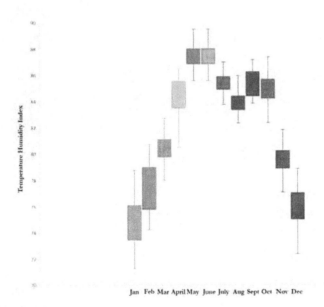

Fig. 8 Temperature humidity index using maximum temperature at Ilela, Sokoto State, as reference point for Sahel ecoclimate zone. According to the information in Table 1

scientists (28.2%) that are dealing with livestock production in the various agro-ecological zones covered in this chapter. About 10% of the respondents were dealing in buying and selling of livestock and poultry. Half the number of the stakeholders interviewed about the interrelationships between climate change and livestock

Table 1 The temperature humidity index chart

Temp F	C	25	30	35	40	45	50	55	60	65	70	75	80	85	90	95	100	
77	25.0						72	72	73	73	74	74	75	75	76	76	77	MILD
78	25.6		NO STRESS			72	73	73	74	74	75	75	76	76	77	77	77	STRESS
79	26.1				72	76	73	74	74	75	76	76	77	77	78	78	79	
80	26.7		72	72	73	76	74	74	75	76	76	77	78	78	79	79	80	
81	27.2	72	72	73	73	74	75	75	76	77	77	78	78	79	80	80	81	
82	27.8	72	73	73	74	75	75	76	77	77	78	79	79	80	81	81	82	
83	28.3	73	73	74	74	75	76	77	78	78	79	80	80	81	82	82	83	SEVERE
84	28.9	73	74	75	75	76	77	78	78	79	80	80	81	82	83	83	84	STRESS
85	29.4	74	75	75	76	77	78	79	79	80	81	81	82	83	84	84	85	
86	30.0	74	75	76	77	78	78	79	80	81	81	82	83	84	84	85	86	
87	30.6	75	76	77	77	78	79	80	81	81	82	83	86	85	85	86	87	
88	31.1	75	76	77	78	79	80	81	81	82	83	84	85	86	86	87	88	
89	31.7	76	77	78	79	80	81	82	83	84	85	86	86	87	88	89	89	
90	32.2	77	78	79	79	80	81	82	83	84	85	86	86	87	88	89	90	
91	32.8	77	78	79	80	81	82	83	84	85	86	86	87	88	89	90	91	
92	33.3	78	79	80	81	82	83	84	85	85	86	87	88	89	90	91	92	
93	33.9	79	80	80	81	82	83	84	85	86	87	88	89	90	91	92	93	VERY
94	34.4	79	80	81	82	83	84	85	86	87	88	89	90	91	92	93	94	SEVERE
95	35.0	80	81	82	83	84	85	86	87	88	89	90	91	92	93	94	95	STRESS
96	35.6	80	81	82	83	85	86	87	88	89	90	91	92	93	94	95	96	
97	36.1	81	82	83	84	85	86	87	88	89	91	92	93	94	95	96	97	
98	36.7	82	83	84	85	86	87	88	89	90	91	93	94	95	96	97	98	
99	37.2	82	83	84	85	87	88	89	90	91	92	93	94	96	97	98	99	
100	37.8	83	84	85	86	87	88	90	91	92	93	94	95	97	98	99	100	
101	38.3	83	86	86	87	88	89	90	92	93	96	95	96	97	99	100	101	
102	38.9	86	85	86	87	89	90	91	92	96	96	96	97	96	99	101	102	
103	39.4	86	86	87	88	89	91	92	94	95	96	97	98	100	101	102	103	
104	40.0	85	86	88	88	90	91	93	94	95	96	97	99	100	101	103	104	
105	40.6	86	87	88	89	91	92	93	96	96	97	98	99	100	101	104	105	DEAD
106	41.1	86	88	89	90	91	93	94	95	97	98	99	101	102	103	105	106	CATTLE
107	41.7	87	88	89	91	92	94	95	96	98	99	101	102	103	105	106	107	
108	42.2	87	89	90	92	93	94	96	97	98	100	101	102	104	105	106	108	
109	42.8	88	89	91	92	94	95	96	98	99	101	102	103	105	106	107	109	
110	43.3	88	90	91	92	94	96	97	98	100	101	102	104	105	106	108	110	
111	43.9	89	91	93	94	95	96	98	99	101	102	103	105	106	107	109	111	

Source Dr. Frank Wiersma (1990). Department of Agricultural Engineering, University of Arizona, Tucson. Downloaded from http://www.veterinaryhandbook.com.au

production preferred intensive management system of production. This can be explained because more than 40% of them are into commercial poultry production. Semi-intensive is a system of choice for ruminant animal production and it is preferred by 40.8% of the stakeholders interviewed in this chapter. Umunna et al. (2014) reported 56.3% of small ruminant producers rearing their stock through semi-intensive system.

The distribution of respondents was also shown in Table 3. The largest proportion of respondents (32.5%) was from Derived Savannah zone of Nigeria. This is commensurate to the very large land area of this zone when compared with some of the other zones (Fig. 1). The least population of respondents (6.7%) was from the Sahel zone. The Sahel zone in Nigeria is found at the uppermost portion of the country. Suleiman (2017) described the Sahel region of Africa as a 3,860-kilometer

Table 3 Features of Savannah and *Sudano-Sahelian* zones being experienced by respondents

Feature	Sahel N=22	Sudan N= 57	Northern Guinea Savannah N = 61	Southern Guinea Savannah N = 80	Derived Savannah N = 106	Total N = 326
Seasonal variation in availability of natural forage	22 (100.0)	57 (100.0)	58 (95.1)	32 (40.0)	48 (45.3)	217 (66.6)
Extreme high temperatures during dry season	17 (77.3)	38 (66.7)	41 (67.2)	55 (68.8)	57 (53.8)	208 (63.8)
Low temperature during Harmattan	21 (95.5)	46 (80.7)	24 (39.3)	18 (22.5)	18 (17.0)	127 (39.0)
Low precipitation	19 (86.4)	39 (68.4)	17 (27.9)	20 (25.0)	23 (21.7)	118 (36.2)
Desert encroachment	13 (59.1)	26 (45.6)	13 (21.3)	22 (27.5)	19 (17.9)	93 (28.5)
Sunshine hours more than 12 hours	18 (81.8)	42 (73.7)	14 (23.0)	13 (16.3)	5 (4.7)	92 (28.2)
Abundance of grasses and other fodder crops	6 (27.3)	5 (8.8)	16 (26.2)	19 (23.8)	26 (24.5)	72 (22.1)
Low to moderate relative humidity	4 (18.2)	8 (14.0)	11 (18.0)	12 (15.0)	10 (9.4)	45 (13.8)
Factors responsible for large population of livestock in Savannah and *Sudano-Sahelian* zones of Nigeria						
Abundance of grasses, legumes and other fodder crops	11 (50.0)	33 (57.9)	36 (5.9)	48 (60.0)	67 (63.2)	195 (59.8)
Large expanse of grassland	18 (81.8)	22 (38.6)	29 (47.5)	41 (51.3)	43 (40.6)	153 (46.9)
Low infestation of pathogens during wet season	6 (27.3)	27 (47.4)	15 (24.6)	32 (40.0)	30 (28.3)	110 (33.7)
Low infestation of pathogens during dry season	10 (45.5)	17 (29.8)	22 (36.1)	34 (42.5)	22 (20.8)	105 (32.2)
Mostly flat plane topography	5 (22.7)	11 (19.3)	16 (26.2)	20 (25.0)	21 (19.8)	73 (22.4)

N is the number of respondents; values in parenthesis are the percentages of their respective frequencies

arc-like land mass lying to the immediate South of the Sahara Desert and stretching East-West across the breadth of the African continent. He further stated that the region stretches from Senegal on the Atlantic coast, through parts of Mauritania, Mali, Burkina Faso, Niger, Nigeria, Chad, and Sudan to Eritrea on the Red Sea coast.

Almost all the respondents (92%) were aware of the concept of climate change and its other attribute of global warming. Very high awareness level of climate change (88%) was reported by Adebayo and Oruonye (2012) among farmers in Northern Taraba State.

The features that best describe Savannah and *Sudano-Sahelian* zones of Nigeria were presented in Table 3. Seasonal variation in availability of natural forage was reported by all the respondents interviewed in Sahel and Sudan zones. About 95% of the respondents in Northern Guinea Savannah zones corroborated the scarcity or non-availability of natural forages during the dry seasons. Life-threatening high temperature during dry season was also reported as 63.8% by 326 respondents. Low temperatures during *Harmattan* period were reported as 95.5%, 80.7%, 39.3%, 22.5%, and 17.0% by respondents from Sahel, Sudan, Northern Guinea Savannah, Southern Guinea Savannah, and Derived Savannah, respectively.

The *Harmattan* is a season in the West African subcontinent starting from November to mid-March. The season is highly dependent on air pressure variability in the Mediterranean area. The Harmattan period is dust laden and also characterized by low temperatures (Schwanghart and Schutt 2008). In Sahelian parts of Africa, Aeolian dust transport is made possible by several wind systems (Jäkel 2004; Engelstaedter et al. 2006). One of the wind system is *Harmattan* (Schwanghart and Schutt 2008).

Low precipitation was also reported in Table 3. The proportions of the respondents that stated low precipitation as a prominent feature of the climate system were highest for Sahel (86.4%) and lowest for Derived Savannah (21.7%). This is an indication that there is more aridity in the Sahel and less in the Derived Savannah. Variability in Sahel rainfall is inextricably connected with the variability of the atmospheric circulation. Annual mean rainfall in the Sahel of Nigeria is less than 200 mm (Biasutti 2019). The author opined that across the zones, abundance or scarcity of rainfall and its distribution over the rainy season and the associated maximum temperature extremes determines the success or failure of farming system with its antecedent effects on livestock production. Desert encroachments were reported as a feature of Sahel (59.1%) and Sudan (45.6%) zones. Nigeria is faced with rapid desert encroachment affecting 15 states in the North. Most of the States covered in this chapter were described as desertification frontline States by Olagunju (2015).

Livestock Population in Nigeria

The total population of cattle in Nigeria was 20,407,607 in 2019 as against 20,231,589 in 2018. The distribution of cattle in States within Nigeria was illustrated through Fig. 9. Zamfara tops the list of States with 3,432,486 heads of cattle. The goat population in Nigeria was totaled at 46,757,458 in 2019. The highest population of goats (5,488,904) in 2019 was recorded in Katsina State (Fig. 10). Like as it is for cattle, Zamfara State tops the list of states for sheep production with the population size of 7,314,023 sheep (Fig. 11). These populations were reported in the Executive summary of Annual Performance Survey of National Agricultural Extension and Research Liaison Services in Nigeria (NAERLS 2019).

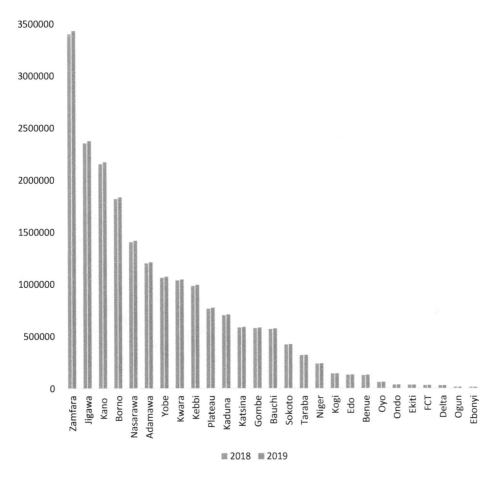

Fig. 9 Cattle population in Nigeria. (Source: Federal Department of Animal Production and Husbandry Services, FMARD, Abuja (Reported by NAERLS 2019))

The total populations of donkeys in Nigeria were 978,402 and 979,380 for 2018 and 2019, respectively (NAERLS 2019). The beast of burden (donkey), a very resilient animal is found mostly in about 11 states of the country (all within Sahel, Sudan, and Northern Guinea Savannah Zone of Nigeria) with the highest population found in Zamfara State (331,641) in 2019. Other states with prominent populations of donkeys in 2019 were Sokoto (153,657), Borno (143,707), Kano (135,962), Kebbi (82,870), Jigawa (25,135), and Gombe (14,241). Some other states like Bauchi and Yobe had populations of donkeys that are less than 1,500.

Camel is another livestock used as beast of burden in Nigeria. The total populations of camels in the country were 279,956 and 280,235 for 2018 and 2019, respectively. Almost half of all the camel population in Nigeria was found in Kano State with 128,104 heads of camel. Other states with some populations of camel in 2019 were Sokoto (60,346), Kebbi (50,483), Jigawa (12,851), Katsina (9,581), Bauchi (9,475), Niger (3,270), and Yobe (501). It was of note that the rate of increase in population of camel and donkey is very negligible. These animals (camel and donkey) are reported to

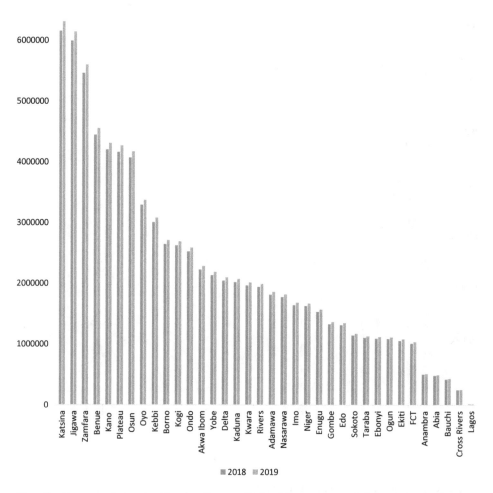

Fig. 10 Goat population in Nigeria. (Source: Federal Department of Animal Production and Husbandry Services, FMARD, Abuja (Reported by NAERLS 2019))

be dwindling in number as there is increased consumption and less production, therefore ways of increasing the population of this animal should be scientifically exploited to avoid the extinction of the species (Nelson et al. 2015).

The possible factors responsible for large population of livestock in the Savannah and *Sudano – Sahelian* zones of Nigeria were presented in Table 3. On the top of the list of such factors is the abundance of grasses, legumes, and other fodder crops as indicated by 59.8% of the respondents. Large expanse of grassland was also said to be a prominent factor enabling large population of livestock on the semiarid zone of Sahel, Sudan, and the Guinea Savannahs. Other factors being reported in favor of the large population of livestock in the zones being considered in this chapter were low infestation of pathogens during wet and dry seasons with 33.7% and 32.2%, respectively. About 22% of the respondents stated that the flat plane topography in the zones might have contributed to the enormous populations of livestock being found in the zones. Lawal-Adebowale (2012) stated that the concentration of

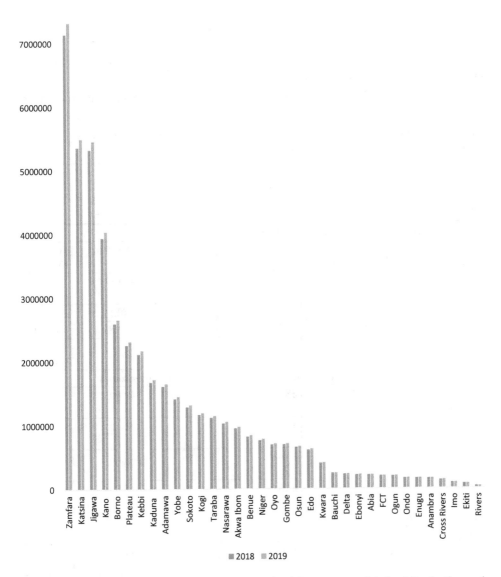

Fig. 11 Sheep population in Nigeria. (Source: Federal Department of Animal Production and Husbandry Services, FMARD, Abuja (Reported by NAERLS 2019))

Nigeria's livestock-base in the northern region is most likely to have been influenced by the ecological condition of the region which is characterized by low rainfall duration, lighter sandy soils, and longer dry season. This submission was predicated by the fact that drier tropics or semi-arid regions are more favorable to the ruminants. However, concerted efforts need to be made at retaining the large population of livestock in these regions (Savannah and *Sudano-Sahelian*) because livestock production will be possibly limited in the future by climate variability as animal's water consumption is expected to increase. There will be more demand for agricultural lands because of increase due to need for 70% growth in production, and food security concern since about one-third of the global cereal harvest will be needed for livestock feed (Rojas-Downing et al. 2017).

Table 4 showed the stakeholders perception of the effect of changes on climatic elements of livestock production in the Savannah and *Sudano-Sahelian* zone of Nigeria. About 77% of all the respondents agreed to the fact that changes in climatic

Table 4 Stakeholders' perception of effect of changes in climatic elements on livestock production in the Savannah and Sudano-Sahelian zones of Nigeria

	Sahel N = 22	Sudan N = 57	Northern Guinea Savannah N = 61	Southern Guinea Savannah N = 80	Derived Savannah N = 106	Total N = 326
Changes in climatic elements affect livestock production in the zones						
Strongly agree	10 (45.5)	11 (19.3)	19 (31.1)	24 (30.0)	38 (35.8)	102 (31.3)
Agree	7 (31.8)	31 (54.4)	23 (37.7)	45 (56.2)	42 (39.6)	148 (45.4)
Neutral	0 (0.0)	4 (7.0)	8 (13.1)	4 (0.5)	9 (8.5)	25 (7.7)
Disagree	1 (4.5)	4 (7.0)	2 (3.3)	2 (2.5)	4 (3.8)	13 (4.0)
Strongly disagree	4 (1.2)	7 (2.1)	9 (2.8)	5 (6.2)	13 (12.3)	38 (11.7)
Perception about climatic elements that have the most variation in the zones						
Atmospheric temperature	9 (40.9)	31 (54.4)	33 (54.1)	41 (51.2)	49 (46.2)	163 (50.0)
Rainfall	10	21	19	28	42	120 (36.8)
Sunshine hour	2 (9.1)	3 (5.3)	3 (4.9)	7 (8.8)	4 (3.8)	19 (5.8)
Relative humidity	0 (0.0)	2 (3.5)	4 (6.6)	2 (2.5)	8 (7.5)	16 (4.9)
Atmospheric pressure	1 (4.5)	0 (0.0)	1 (1.6)	2 (2.5)	3 (2.8)	7 (2.1)
Solar radiation	0 (0.0)	0 (0.0)	1 (0.3)	0 (0.0)	0 (0.0)	1 (0.3)
Perception about the climatic elements that are capable of affecting livestock productivity when their variations are in the extreme						
Atmospheric temperature	17 (77.3)	39 (68.4)	48 (78.7)	54 (67.5)	69 (65.1)	227 (69.6)
Relative humidity	2 (9.1)	12 (21.1)	14 (23.0)	17 (21.3)	35 (33.0)	80 (24.5)
Solar radiation	3 (13.6)	13 (22.8)	9 (14.8)	15 (18.8)	18 (20.0)	58 (17.8)
Sunshine hours	4 (18.2)	7 (12.3)	12 (19.7)	19 (23.8)	21 (19.8)	63 (19.3
Greenhouse gases	2 (9.1)	4 (7.0)	9 (14.8)	4 (5.0)	7 (6.6)	26 (10.0)

N is the number of respondents; values in parenthesis are the percentages of their respective frequencies

elements affect livestock productivity. Kebede (2016) related the foremost reaction of animals under thermal weather as increase in respiration rate, rectal temperature, and heart rate. He further stated that the anticipated rise in temperature due to climate change is likely to aggravate the heat stress in livestock, adversely affecting their productive and reproductive performance and even death in extreme cases. The respondents observed that the climatic elements with most variations are atmospheric temperature (50.0%), rainfall (36.8%), and sunshine hour (5.8%). The climatic element with the least variation as being reported by the respondents is solar radiation (0.3%). Atmospheric temperature was also implicated by 69.9% of the respondents as a climatic element with the most debilitating effect on livestock when its variation is in the extreme. This was followed by relative humidity with 24.5% of the respondents stating that its effect can really affect livestock productivity.

Adaptive Measures Against the Effect of Climate Change on Livestock Production

Useable adaptive measures toward reducing the effects of climate change on live-stock production are presented in Table 5. About 55% of the respondents agreed that the use of adaptive measures in alleviating the effect of climate change on livestock is capable of reducing its debilitating effect on livestock. The rest of the respondents, about 45%, were either neutral or disagreed with the fact that adaptive measures can mitigate the effect of climate change. It will be necessary to educate those that disagree on this very important fact. To guide the evolution of livestock production systems under the increase of temperature and extreme events, better information is needed regarding biophysical and social vulnerability, and this must be integrated with agriculture and livestock components (Nardone et al. 2010). The specific adaptive measures used by livestock farmers in the study locations are shown in Table 5 as well. At the top of the adaptive features of choice by respondents is provision of housing facilities for animals which was indicated by about 60% of the respondents. Provision of abundant water and supplements feeding were also indicated as adaptive measures by 45.4% and 44.2% of the respondents, respectively. Planting of tress to provide shades for livestock was of great interest because of the sustainable effect of this adaptive measure to livestock production. Trees are known to absorb carbon dioxide produced by man and animals that is apart from their primary function of shades as intended by livestock farmers. Development of super-absorbent fake leaves was proposed by scientists (Vince 2012) as a means of modulating the global temperature. This method was proposed as capable of removal of greenhouse gas from the atmosphere. The benefits of the introduction of artificial plants will be centered on geoengineering the planet which will be beyond its cooling effects.

Timely control of internal and external parasites was a choice of adaptive measure by a third of the respondents (31.3%). This is expected to eliminate the stress on health status of the animals which will go a long way in stabilizing the internal physiological equilibrium of the animals. If properly done, the animals will have enough energy to

Table 5 Useable adaptive measures toward reducing the effect of climate change on livestock production in Savannah and Sudano-Sahelian zones of Nigeria

Feature	Sahel N = 22	Sudan N = 57	Northern Guinea Savannah N = 61	Southern Guinea Savannah N = 80	Derived Savannah N = 106	Total N = 326
Stakeholders perceptions about reducing the effect of climate change on livestock production through the use of adaptive measures						
Strongly agree	2 (9.1)	5 (8.8)	0 (0.0)	0 (0.0)	0 (0.0)	7 (2.1)
Agree	10 (45.5)	31 (54.4)	28 (45.9)	45 (56.3)	58 (54.7)	172 (52.8)
Neutral	4 (18.2)	6 (10.5)	3 (4.9)	6 (7.5)	5 (4.7)	24 (7.4)
Disagree	0 (0.0)	4 (7.0)	1 (1.6)	0 (0.0)	5 (4.7)	10 (3.1)
Strongly disagree	6 (27.3)	11 (19.3)	29 (47.5)	29 (36.3)	38 (35.8)	113 (34.7)
Adaptive measures used by livestock production stakeholders						
Provision of housing for animals	17 (77.3)	41 (71.9)	35 (57.4)	51 (63.8)	50 (47.2)	194 (59.5)
Frequent cleaning of animal houses	6 (27.3)	17 (29.8)	24 (39.3)	24 (30.0)	31 (29.2)	102 (31.3
Provision of supplement feeding	9 (40.9)	31 (54.4)	24 (39.3)	41 (51.3	39 (36.8)	144 (44.2)
Provision of water in abundance	12 (54.5)	31 (54.4)	27 (44.3)	38 (47.5)	39 (36.8)	148 (45.4)
Timely control of internal and external parasites	7 (31.8)	17 (29.8)	19 (31.1)	28 (35.0)	32 (30.2)	103 (31.6)
Storage of excess feed materials	9 (40.9)	18 (31.6)	24 (39.3)	31 (38.8)	24 (38.8)	106 (32.5)
Cultivation of drought tolerant varieties of forage crops	4 (18.2)	18 (31.6)	19 (31.1)	27 (33.8)	23 (21.7)	91 (27.9)
Feeding of livestock with crop residues	8 (36.4)	16 (28.1)	20 (32.8)	23 (28.8)	25 (23.6)	92 (28.2)
Making of multi-nutrient blocks	3 (13.6)	8 (14.0)	10 (16.4)	14 (17.5)	13 (12.3)	48 (14.7)
Feeding of livestock with multi-nutrient blocks	7 (31.8)	10 (17.5)	7 (11.5)	12 (15.0)	11 (10.4)	47 (14.4)
Seasonal migration of animals	3 (13.6)	2 (3.5)	12 (19.7)	6 (7.5)	17 (16.0)	40 (12.3)

(continued)

Table 5 (continued)

Feature	Sahel $N = 22$	Sudan $N = 57$	Northern Guinea Savannah $N = 61$	Southern Guinea Savannah $N = 80$	Derived Savannah $N = 106$	Total $N = 326$
Irrigation of pasture during dry season	5 (22.7)	7 (12.3)	13 (21.3)	21 (26.3)	8 (7.5)	54 (16.6)
Establishment of ranch	4 (18.2)	12 (21.1)	11 (18.0)	21 (26.3)	20 (18.9)	68 (20.9)
Planting of trees to provide shades for livestock	11 (50)	26 (45.6)	29 (47.5)	37 (46.3)	30 (28.3)	133 (40.8)
Storage of crop residues obtainable during crop harvest	7 (31.8)	14 (24.6)	20 (32.8)	23 (28.8)	22 (20.8)	86 (26.4)

N is the number of respondents; values in parenthesis are the percentages of their respective frequencies

combat stress from the environment. Storage of excess feed, especially during harvest, was stated as an adaptive measure by 32.5% of the respondents. This adaptive measure can be linked with another one that was also stated by the stakeholders, storage of crop residues obtainable during harvest (26.4%). These two measures are some of the important components of crop-livestock integration systems as discussed by Iyiola-Tunji et al. (2015). Feeding livestock with crop residues in a well-planned basis on the nutrient requirements and biomass needs of these animals will ensure adequate usage of the crop residues. Establishment of ranch, irrigation of pasture during dry season, making of multi-nutrient blocks, feeding of livestock with multi-nutrient blocks and seasonal migration of animals were of the other adaptive measures being carried out to combat the effect of climate change as reported by substantial proportion of the respondents. Integrating livestock and crop production will serve as a form of conservation, which will enable shifting from the traditional systems which is focused exclusively on livestock or crop to a new approach which sustainably combines both. Agroforestry (establishing trees alongside crops and pastures in a mix) as a land management approach can help maintain the balance between agricultural production, environmental protection, and carbon sequestration to offset emissions from the sector. Agroforestry may increase productivity and improve quality of air, soil, and water, biodiversity, pests and diseases, and improves nutrient cycling (Jose 2009; Smith et al. 2012).

Contribution of Livestock Production Activities Toward Climate Change

Table 6 showed the contribution of livestock production activities toward climate change. A lot of the stakeholders interviewed (62.3%) were aware of the

Table 6 Contribution of livestock production activities toward climate change

	Sahel $N = 22$	Sudan $N = 57$	Northern Guinea Savannah $N = 61$	Southern Guinea Savannah $N = 80$	Derived Savannah $N = 106$	Total $N = 326$
Stakeholder awareness of contribution of livestock production activities to changes in climate and global warming						
Yes	16 (72.7)	32 (56.1)	43 (70.5)	44 (55.0)	68 (64.2)	203 (62.3)
No	2 (9.1)	11 (19.3)	1 (1.6)	11 (13.8)	8 (7.5)	33 (10.1)
Maybe	1 (4.5)	2 (3.5)	9 (14.8)	6 (7.5)	7 (6.6)	25 (7.7)
Livestock generates substantial proportions of global greenhouse gas emission that are very bad for the environment						
Yes	14 (63.6)	28 (49.1)	35 (57.4)	35 (43.8)	62 (58.5)	174 (53.4)
No	5 (22.7)	16 (28.1)	11 (18.0)	33 (41.3)	24 (22.6)	89 (27.3)
Maybe	3 (13.6)	11 (19.3)	15 (24.6)	12 (15.0)	19 (17.9)	60 (18.4)
Livestock and their by-products account for several millions tons of carbon dioxide production per year						
Yes	12 (54.5)	22 (38.6)	25 (40.9)	28 (35.0)	57 (53.8)	114 (35.0)
No	7 (31.8)	21 (36.8)	13 (21.3)	27 (33.8)	25 (23.6)	93 (28.5)
Maybe	3 (13.6)	14 (24.6)	23 (37.7)	25 (31.3	23 (21.7)	88 (27.0)
Extensive system of livestock production plays a critical role in land degradation, climate change, water, and biodiversity loss						
Yes	14 (63.6)	29 (5.1)	44 (72.1)	67 (83.8)	86 (81.1)	240 (73.6)
No	2 (9.1)	11 (19.3)	5 (8.2)	6 (7.5)	6 (5.7)	30 (9.2)
Maybe	5 (22.7)	14 (24.6)	9 (14.8)	7 (8.8)	14 (13.2)	49 (15.0)
Economic, social, health, and environmental perspectives will be critical to solving the problems surrounding livestock production						
Strongly agree	6 (27.3)	18 (31.6)	26 (42.6)	32 (40.0)	48 (45.3)	130 (39.9)
Agree	12 (54.5)	27 (47.4)	29 (47.5)	41 (51.3)	47 (44.3)	156 (47.9)
Neutral	3 (13.6)	5 (8.8)	0 (0.0)	5 (6.3)	7 (6.6)	20 (6.1)
Disagree	0 (0.0)	5 (8.8)	5 (8.2)	2 (2.5)	4 (3.8)	16 (4.9)
Strongly disagree	1 (4.5)	2 (3.5)	0 (0.0)	0 (0.0)	0 (0.0)	3 (0.9)

N is the number of respondents; values in parenthesis are the percentages of their respective frequencies

contribution of livestock production to climate change. Generations of substantial proportions of global greenhouse gases that are very bad for the environment were on the knowledge of more than half of the respondents (53.4%). Just about the third (35%) of the respondents were aware that livestock and their by-products account for several million tons of carbon dioxide production per year. Very large proportions (73.6%) of respondents were aware that extensive system of livestock production plays a critical role in land degradation, climate change, water, and biodiversity loss. About 90% of the respondents however believed that economic, social, health, and environment perspectives are critical to solving the problems of the contributions of livestock production to climate change and global warning. In 2006, an FAO publication entitled "Livestock's long shadow – Environmental issues and options" indicated that the influence of livestock on the environment was much greater than it was considered. This provided detailed perspectives on the impact of livestock on water, biodiversity, and climate change. The issue on climate change and 18% estimated contribution of livestock to overall GHG emissions is the concern that attracted the most attention. The FAO (2006) estimated 18% anthropogenic GHG emissions from livestock industry is disapproved by Goodland and Anhang (2009) who noted that the figure under-tallies emissions from certain production activities, underestimates demand, and absolutely omits some categories of emissions. They estimated that livestock production is contributing about 51% of anthropogenic GHG emissions. Goodland and Anhang (2009) revealed that CO_2 from livestock respiration was ignored as a source of the GHGs from the FAO study (2006). Both manure and enteric fermentation contribute some 80% of methane emissions from agricultural activities and about 30–40% of the overall anthropogenic methane emissions (FAO 2006). The 62–89% of greenhouse emission recorded in this chapter was similar to the findings of FAO (2006). Similarly, there is an increasing awareness within the policy and research communities that fast growth in consumption and production of livestock commodities is contributing to variety of environmental problems. The main notable issue is livestock's significant contribution to anthropogenic emissions. Majority of the revenue is generated by pigs, chickens, sheep, goats, beef, and dairy cattle. These five species of livestock generate 92% of the overall revenue from livestock in Africa. In most rural communities, livestock is the only property of the poor, but it is highly susceptible to climate changes and extremes (Easterling and Aggarwal 2007; FAO 2007; Calvasa et al. 2009). The influence of climate change is anticipated to increase the susceptibility of livestock industry and reinforce current factors that are having impact on livestock farming systems (Gill and Smith 2008). The overall GHG emissions from livestock supply chains are approximately 7.1 gigatons CO_2-equivalent annually for the 2005 reference point forming about 14.5 % of all emissions induced by humans (IPCC 2007a). About 44 % of the livestock industry emissions are in the form of CH_4. Nitrous oxide and carbon dioxide represent 29% and 27%, respectively. Livestock supply chains emit 9.2 gigatons CO_2-eq of CO_2 annually or 5% of anthropogenic CO_2 emissions (IPCC 2007b). According to IPCC (2007b), 44% of anthropogenic CH_4 emissions or 3.1 gigatons CO_2-eq of CH_4 every year and 53% of anthropogenic N_2O emissions or 2 gigatons CO_2-eq of N_2O are produced annually. Similar results were observed in this chapter, which reported that livestock products account for 88–93% (Table 6) of the carbon dioxide production per year.

Adaptive Measures Toward Mitigation of Effect of Climate Change on Livestock

An adaptation such as the modification of production and management systems involves diversification of livestock animals and crops, integration of livestock systems with forestry and crop production, and changing the timing and locations of farm operations (IFAD 2010). Diversification of livestock and crop varieties can increase drought and heat wave tolerance, and may increase livestock production when animals are exposed to temperature and precipitation stresses. In addition, this diversity of crops and livestock animals is effective in fighting against climate change-related diseases and pest outbreaks (Kurukulasuriya and Rosenthal 2003; Batima et al. 2005; IFAD 2010). Changes in breeding strategies can help animals increase their tolerance to heat stress and diseases and improve their reproduction and growth development (Rowlinson et al. 2008; Henry et al. 2012). Adjusting animal diets can also be used as a mitigation measure, by changing the volume and composition of manure. GHG emissions can be reduced by balancing dietary proteins and feed supplements. If protein intake is reduced, the nitrogen excreted by animals can also be reduced. Supplements such as tannins are also known to have the potential to reduce emissions. Tannins are able to displace the nitrogen excretion from urine to feces to produce an overall reduction in emissions (Hess et al. 2006; Dickie et al. 2014). Some of the adaptable technologies for reducing the effect of livestock production activities on climate change and vice versa are also presented in Table 7 and discussions on each of them are presented below.

Proper Livestock Health Management and Welfare

On the top of the list of technologies as dictated by the respondents (63.2%) is proper livestock health management and welfare. Reducing greenhouse gas (GHG) emissions may seem like extra work that can hurt business, but in reality, best management practices for reducing GHG emissions can be economical (Lindgren 2019). Animals that are maintained in optimum health conditions and given adequate welfare will have improved production efficiency and reduction of methane production from digestion of feeds.

Adequate Waste Management and Utilization

Almost equally important technology is adequate waste management and utilization as proposed by 59.2% of the respondents. The major contribution to greenhouse gas emissions is methane (CH_4) from ruminant animals through belching when the animals digest their feeds (Plate I). The other sources of the deleterious gases are from fecal waste excretion and storage. Adequate waste management and utilization is capable of reducing the quantity of the greenhouse gases

Table 7 Adaptable technologies for reducing the effect of livestock production activities on climate change

Adaptable technologies	Sahel $N = 22$	Sudan $N = 57$	Northern Guinea Savannah $N = 61$	Southern Guinea Savannah $N = 80$	Derived Savannah $N = 106$	Total $N = 326$
Proper livestock health management and welfare	13 (59.1)	40 (70.2)	34 (55.7)	55 (68.8)	64 (60.4)	206 (63.2)
Adequate waste management and utilization	13 (59.1)	32 (56.1)	40 (65.6)	43 (53.8)	65 (61.3)	193 (59.2)
Crop-livestock integration system	9 (40.9)	28 (49.1)	32 (52.5)	36 (45.0)	59 (55.7)	164 (50.3)
Breeding for more productive animals	12 (54.5)	31 (54.4)	28 (45.9)	43 (53.8)	49 (46.2)	163 (5.0)
Use of methane reducing feed additives	9 (40.9)	21 (36.8)	21 (34.4)	13 (16.3)	22 (20.8)	86 (26.4)
Ranching	8 (36.4)	15 (26.3)	18 (29.5)	17 (21.3)	26 (24.5)	84 (25.8)

N is the number of respondents; values in parenthesis are the percentages of their respective frequencies

Plate I Greenhouse gas emissions from cattle production. (Source: Lindgren (2019))

emissions. Livestock farmers in the Sahel, Sudan, and the Guinea Savannah zones of Nigeria use the fecal waste as organic fertilizers for crop production. There were occasions where the litter materials from poultry production are fed to cattle (Lamidi 2005).

Crop-Livestock Integration Systems

A lot of the effect of livestock production on climate change can be eliminated if the farmers can engage in crop-livestock integration systems. About half of the respondents (50.3%) agreed to this fact. Ickowicz et al. (2012) presented three variants of CLIS in

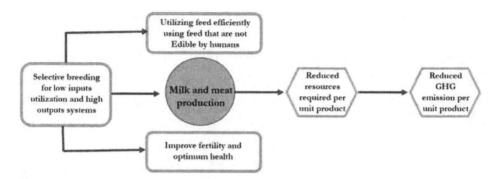

Fig. 12 Production efficiencies using management that can reduce GHG emissions beginning with selective breeding of a genotype for a particular system. (Adapted from Bell et al. (2012); modified by Iyiola-Tunji, A.O.)

arid and semiarid areas: (i) livestock only grazing systems, (ii) rainfed mixed crop-livestock systems, and (iii) irrigated mixed crop-livestock systems. CLIS combine cereal crops (mainly millet, cowpea, sorghum, cotton and groundnut) and majorly ruminant animal production activities in different proportions. Crop-livestock integration systems (CLIS) enable recycling of products and wastes between crop production and livestock production. These methods are capable of increasing feed resources availability during the dry season and also replenish the soil for crop production through the use of fecal wastes from livestock. The major engagement of agro-pastoralists in Nigeria involves CLIS in a way though biomass inputs and outputs recycling are not scientifically calculated by the farmers (Iyiola-Tunji et al. 2017).

Breeding for More Productive Animals

Breeding for more productive animals was suggested by 50% of the respondents as an adaptive measure for reduction of greenhouse gas emissions. Selective breeding that is aimed at improving production efficiencies had been reported to result into increase productivity and gross efficiency by optimize the cost of production and reduce the number of animals that are needed to produce the same quantity of products (Bell et al. 2012). Reports from van de Haar and St. Pierr (2006) and Chagunda et al. (2009) related that more energy-efficient animals produce less waste in the form of methane and nitrogen excretion per unit product. The path toward reduced emission of greenhouse gases through selective breeding is depicted in Fig. 12. Animals that are selectively bred to utilize low inputs and give high outputs are expected to produce milk and meat (as the case may be) efficiently. The quantity of GHG emissions will be reduced once the number of animals put into productive is reduced.

Use of Methane-Reducing Feed Additives

The use of methane reducing feed additives was stated by 26.4% of the respondents as being capable of reducing the effect of livestock production activities on GHG emissions. Kataria (2015) observed that the practice of using feed additives to mitigate enteric methane production is more prominent in developed countries of the world where ruminant livestock are kept in well-managed production systems and generally fed diets that are very high in digestibility and nutrients. The results of this practice according to the author are an efficient production (milk or meat) relative to the amount of methane emitted. Klop (2016) expressed the advantage of using feed additives to mitigate GHG emissions as they are supplied in such amounts that the basal diet composition will not be largely affected by the feed additives (Klop 2016). Methane-reducing feed additives and supplements inhibit methanogens in the rumen, and subsequently reduce enteric methane emissions (Curnow 2019). Methane-reducing feed additives and supplements can be synthetic chemicals, natural supplements and compounds, such as tannins, and seaweed fats and oils (Curnow 2019). van Zijderveld et al. (2010) had experimented with lauric acid, myristic acid, linseed oil, and calcium fumarate as additives and obtained favorable results in the reduction of GHG emissions. Sunflower oil and monensin offer the greatest reductions in methane without substantial reductions in diet digestibility (Beauchemin and McGinn 2006). It is of note that the practice of using feed additives as an adaptive measure to reduce GHG emissions in developing countries like Nigeria is almost nonexistent.

Ranching

To further reduce livestock's greenhouse gas emissions while continuing to provide meat for a growing world population, beef cattle ranchers are proactively implementing methane-reducing methods to manage manure, improve soil health, and enhance herd efficiency. Ranching will enable farmers to consciously engage in practices that are capable of mitigating the effect of climate change on their livestock and also make attempt at GHG emissions from their livestock.

Pathway of Responses

The dual pathways of responses between climate change and livestock production activities are depicted in Fig. 13. Activities from livestock have very high tendencies to impact negatively on the environment and eventually causing unfavorable variability of climate and its elements, which is indicated by the blue big (fat) arrow that goes away from livestock to the environment and climate. The major component of the activities of livestock that is known to cause injury to the environment as depicted in Fig. 13 is the production of greenhouse gases (shown in an orange box on the right-hand side of the pathway). From the respondents in this study, some adaptive measures were stated as having controlling and mitigating effect at reducing the effect of activities of livestock on the climate and the environment. When these measures such as planting of trees to absorb CO_2, adequate waste management and utilization, feeding of livestock with methane reducing feed additives, and breeding of animals with faster growth rate are effectively deployed, the destruction of the environment will be reduced. Key breeding traits associated with climate change

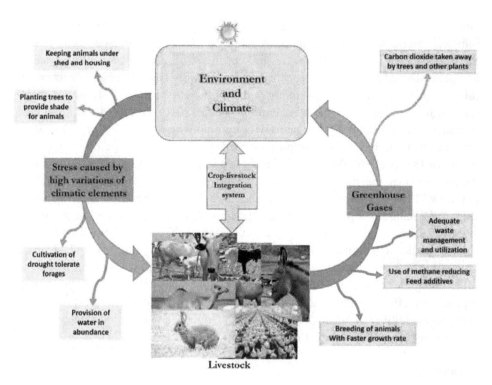

Fig. 13 Dual pathways of responses between climate and livestock

resilience and adaptation include thermal tolerance, low quality feed, high survival rate, disease resistance, good body condition, and animal morphology (Hoffmann 2008; Oseni and Bebe 2008). In general, developing countries have a weak capacity for high-tech breeding programs toward livestock improvement (IFAD 2002). Therefore, programs based on controlled mating methods are likely to be more appropriate. These programs usually do not produce immediate improvements. Improvements are usually not seen for at least one growing season, so a livestock producer must be able to incorporate long-term planning into production management strategies. Such measures could include:

- Identifying and strengthening local breeds that have adapted to local climatic stress and feed sources
- Improving local genetics through cross-breeding with heat and disease tolerant breeds

The environment and climate on the other side of the dual pathway is also known to induce stress on livestock. The respondents in this chapter stated that the components of the pathway that are in yellow boxes are capable of limiting the stress caused by high variations of climatic elements. The concept of crop-livestock integration system is advocated in this chapter as beneficial to livestock and environment in the short and long run.

Predicting Climatic Conditions Using Machine Learning Approach

The ability to forecast climatic conditions is essential for proper planning in live-stock production. Machine learning (ML) approach leverages on past data to predict future events. Three (3) ML model were built to predict the monthly minimum temperature, maximum temperature, and relatively respectively based on information from the previous 11 months.

The methodology adopted is to treat each prediction task as a supervised learning problem. This involves transforming the time series data (Fig. 14) into a feature-target dataset using auto regressive (AR) technique.

The parameter (temp_min or temp_max or relative humidity) to be predicted is set as the target (dependent) variable and in each case be defined by

$$T\min(t)|T\max(t)|RH(t) = f[T\min(t-n), T\max(t-n), RH(t-n)] \qquad (1)$$

t is the prediction date.

t-n denotes the time lags, n is an integer between 1 and 11

Tmin(t), Tmax(t), $RH(t)$ are temperatures and relative humidity to be predicted.

Tmin$(t-n)$, Tmax$(t-n)$, and $RH(t-n)$ are minimum, maximum temperatures, and
relative humidity, respectively, each time lag.

The transformation resulted in a dataset with 445 samples, each with 34 new features. In order to build an ML model, the samples were divided into 361 train

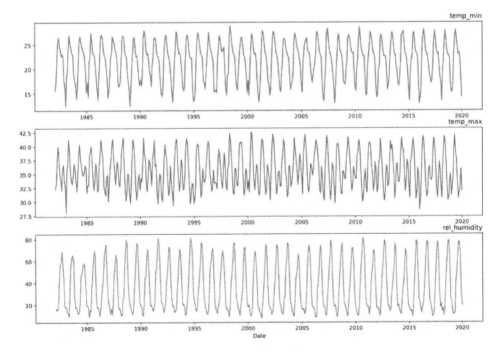

Fig. 14 Time Series of Temperature and Relative Humidity (1982–2019)

(samples from 1982 to 2012) and 84 validations (samples from 2013 to 2019) sets. The Ensemble machine learning methods which are a stack of multiple learning algorithms were used to train our model. The choice of ensemble algorithm is to obtain better predictive performance than could be obtained from any of the constituent learning algorithms. For the three models that were built, the predictive accuracy measured by the R^2 for minimum temperature, maximum temperature, and relative humidity are 0.9353, 0.8772, and 0.9569 respectively. The plots of the actual prediction and the ground truth for minimum and maximum temperatures and relative humidity are shown in Figs. 15, 16, and 17, respectively.

The usefulness of the model developed can be successfully used to predict minimum and maximum temperature as well as relative humidity of Ilela, Sokoto State (representative of Sahel ecoclimate zone). If these predictions are done appropriately, livestock farmers can use the predicted values to calculate temperature humidity index which is indication of level of stress to livestock. Farmers can in essence adjust their management practices accordingly to ensure adequate adaptation in reducing the anticipated stress that may come to their farm animals.

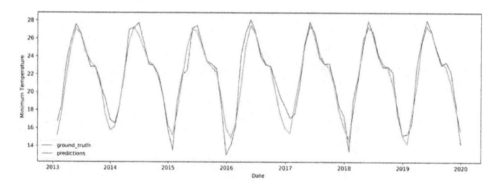

Fig. 15 Plot of predicted and actual values for minimum temperature for Sahel

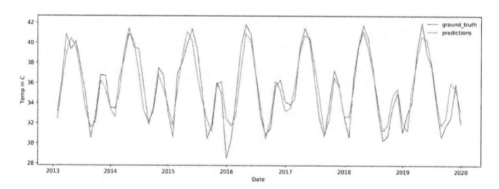

Fig. 16 Plot of predicted and actual values for maximum temperature for Sahel

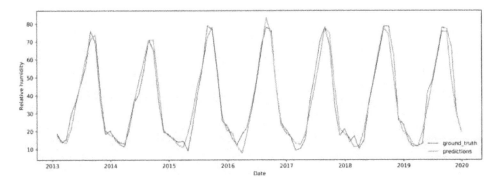

Fig. 17 Plot of predicted and actual values for relative humidity for Sahel

Conclusion and Recommendations

Large proportions of livestock stakeholders in Nigeria are aware of the effect of climate change on livestock production as well as the contributions of livestock production activities to climate change through GHG emissions. About 55% of the respondents agreed that the use of adaptive measures in alleviating the effect of climate change on livestock is capable of reducing its debilitating effect on livestock. The rest of the respondents, about 45%, were either neutral or disagreed with the fact that adaptive measures can mitigate the effect of climate change. It will be necessary to educate those that disagree on this very important fact. About 90% of the respondents however believed that economic, social, health, and environment perspectives are critical to solving the problems of the contributions of livestock production to climate change and global warning. Based on the predictive model developed for temperature and relative humidity in a sample location (Ilela) using Machine Learning in this chapter, there is need for development of a web or standalone application that will be useable by Nigerian farmers, meteorological agencies, and extension organizations as climate fluctuation early warning system. Development of this predictive model needs to be expanded and made functional.

References

AbdulKadir A, Usman MT, Shaba AH (2015) An integrated approach to delineation of the eco-climatic zones in Northern Nigeria. J Ecol Nat Environ 7(9):247–255

Adebayo AA, Oruonye ED (2012) An assessment of the level of farmers awareness and adaptation to climate change in Northern Taraba State, Nigeria. In: Proceedings of the 2012 climate change and ICT conference by Centre for Climate Change and Environmental Research, Osun State University, Osogbo

Adegoke J, Lamptey BL (1999) Intra seasonal variability of summertime precipitation in the Guinea Coastal Region of West Africa'. Paper presented at Cheikh Anta Diop University, Dakar, Senegal, June 1999

Adesogan AT, Havelaar AH, Mckune SL, Eilitta M, Dahl GE (2020) Animal source foods: Sustainability problem or malnutrition and sustainability solution? Perspective matters. Glob Food Secur 25:1–7

Akande A, Costa AC, Mateu J, Henriques R (2017) Geospatial analysis of extreme weather events in Nigeria (1985–2015) using self-organizing maps. Adv Meteorol 2017:11, Article ID 8576150. https://doi.org/10.1155/2017/8576150

Aydinalp C, Cresser MS (2008) The effects of global climate change on agriculture. Am Eurasian J Agric Environ Sci 3(5):672–676

Ayoade JO (2008) Techniques in climatology. Stirling-Horden Publishers Ltd, Ibadan, pp 42–118

Batima P, Bat B, Tserendash L, Bayarbaatar S, Shiirev-Adya S, Tuvaansuren G, Natsagdorj L, Chuluun T (2005) Adaptation to climate change, vol 90. Admon Publishing, Ulaanbaatar

Beauchemin KA, McGinn SM (2006) Effects of various feed additives on the methane emissions from beef cattle. Int Congr Ser 1293:152–155. https://doi.org/10.1016/j.ics.2006.01.042

Bell MJ, Eckard RJ, Pryce JE (2012) Breeding dairy cows to reduce greenhouse gas emissions, livestock production, Khalid Javed, IntechOpen. https://doi.org/10.5772/50395 https://www.intechopen.com/books/livestock-production/breeding-dairy-cows-to-reduce-greenhouse-gas-emissions. Accessed 25 Sept 2020.

Berihulay H, Abied A, He X, Jiang L, Ma Y (2019) Adaptation mechanisms of small ruminants to environmental heat stress. Animals 9:75

Bettencourt EMV, Tilman M, Narciso V, Carvalho MLS, Henriques PDS (2015) The livestock roles in the wellbeing of rural communities of Timor-Leste. Braz J Rural Econ Soc 53(1):S063–S080

Biasutti M (2019) Rainfall trends in the African Sahel: characteristics, processes, and causes. WIREs Clim Change 10:e591. https://doi.org/10.1002/wcc.591

Brown S (2019) How livestock farming affects the environment. https://www.downtoearth.org.in/factsheet/how-livestock-farming-affects-the-environment-64218. Accessed 10 July 2020

Calvasa C, Chuluunbaatar D, Fara K (2009) Livestock thematic papers- tools for project design. International Fund for Agricultural Development (IFAD) publication. Retrieved from www.ifad.org/lrkm/factsheet/cc.pdf. Accessed 10 Dec 2015

Chagunda MGG, Römer DAM, Roberts DJ (2009) Effect of genotype and feeding regime on enteric methane, non-milk nitrogen and performance of dairy cows during the winter feeding period. Livest Sci 122:323–332

Curnow M (2019) Carbon farming: reducing methane emissions from cattle using feed additives. https://www.agric.wa.gov.au/climate-change/carbon-farming-reducing-methane-emissions-cattle-using-feed-additives. Accessed 25 Sept 2020

Dickie A, Streck C, Roe S, Zurek M, Haupt F, Dolginow A (2014) Strategies for mitigating climate change in agriculture: abridged report. Climate focus and california environmental associates, prifadred with the support of the climate and land use Alliance. Report and supplementary materials available at www.agriculturalmitigation.org

Downing TE, Ringuis L, Hulme M, Waughray D (1997) Adapting to climate change in Africa. Mitig Adapt Strat Glob Chang 2:19

Easterling WE, Aggarwal PK (2007) Food, fibre and forest products. In: Parry ML, Canziani OF, Palutikof JP, Van der Linden PJ, Hanson CE (eds) Climate change 2007- impacts, adaptation and vulnerability. Contribution of working group II to the fourth assessment report of the intergovernmental panel on climate change. Cambridge University Press, Cambridge, pp 273–313

Edwards DC, McKee TB (1997) Characteristics of 20th century drought in the United States at multiple time scales. Climatology report no. 97-2, Colorado State University Ft. Collins

Engelstaedter S, Tegen I, Washington R (2006) North African dust emissions and transport. Earth Sci Rev 79:73–100

FAO (1993) Livestock and improvement of pasture, feed and forage. FAO committee on agriculture, 12th session, item 7, 26 April to 4 May 1993, Rome, 19 pp

FAO (2006) In: Steinfeld H, Gerber P J, Wassenaar T, Castel V, Rosales M, de Haan C (eds) Livestock's long shadow – environmental issues and options Food and Agriculture Organization of the United Nations, Rome

FAO (2007) Adaptation to climate change in agriculture, forestry, and fisheries: perspective, framework and priorities. FAO, Rome

FAO (2011) Mapping supply and demand for animal-source foods to 2030, by Robinson TP, Pozzi F. Animal production and health working paper no. 2, Rome

Flachowsky G, Kamphues J (2012) Carbon footprints for food of animal origin: what are the most preferable criteria to measure animal yield? Animals (Basel) 2(2):108–126

Gill M, Smith P (2008) Mitigating climate change: the role of livestock in agriculture. In: Livestock and global change conference proceeding. May 2008, Tunisia

Goodland R, Anhang J (2009) Livestock and climate change. What if the key actors in cli-mate change were pigs, chickens and cows? Worldwatch November/December 2009, Worldwatch Institute, Washington, DC, pp 10–19

Hahn GL (1989) Bioclimatology and livestock housing: theoretical and applied aspects. In: Proccedings of Brazilian workshop on animal bioclimatology. Jaboticabal, p 15

Haider H (2019) Climate change in Nigeria: impacts and responses. K4D (Knowledge, evidence and learning for development) Helpdesk report. https://assets.publishing.service.gov.uk/media/5dcd7a1aed915d0719bf4542/675_Climate_Change_in_Nigeria.pdf

Henry B, Charmley E, Eckard R, Gaughan JB, Hegarty R (2012) Livestock production in a changing climate: adaptation and mitigation research in Australia. Crop Pasture Sci 63:191–202

Hess HD, Tiemann TT, Noto F, Carulla JE, Kreuzer M (2006) Strategic use of tannins as means to limit methane emission from ruminant livestock. In: International conference on Greenhouse gases and animal agriculture, vol. 129, Elsevier International congress series, Zurich, Switzerland, pp 164–167

Hoffmann I (2008) Livestock genetic diversity and climate change adaptation. Livestock and global change conference proceeding. May 2008, Tunisia

Ickowicz A, Ancey V, Corniaux C, Duteurtre G, Poccard-Chappuis R, Touré I, Vall E, Wane A (2012) Crop–livestock production systems in the Sahel- increasing resilience for adaptation to climate change and preserving food security. Building resilience for adaptation to climate change in the agriculture sector. In: Proceedings of a joint FAO/OECD workshop (eds: Meybeck A, Lankoski J, Redfern S, Azzu N, Gitz V). http://www.fao.org/3/i3084e/i3084e.pdf

IFAD (International Fund for Agricultural Development) (2010) Livestock and climate change. http://www.ifad.org/lrkm/events/cops/papers/climate.pdf

IFAD (The International Fund for Agricultural Development) (2002) 'The rural poor' in world poverty report. IFAD, Rome

IPCC (2007a) Summary for policymakers. In: Climate change 2007: impacts, adaptation and vulnerability. Contribution of working group II to the fourth assessment report of the Intergovernmental Panel on Climate Change: Parry ML, Canziani OF, Palutikof JP, Van der Linden PJ, Hanson CE. Cambridge University Press, 1000

IPCC (2007b) Climate change 2007: mitigation. In: Metz B, Davidson O R, Bosch P R, Dave R and Meyer LA (eds) Contribution of working group III to the fourth assessment report of the intergovernmental panel on climate change. Cambridge University Press, Cambridge, UK/New York

Iyiola-Tunji AO (2012) Genetic analysis of growth rate and some reproductive traits of Balami, Uda and Yankasa sheep and their crosses. Unpublished PhD thesis submitted to School of Post Graduate Studies, Ahmadu Bello University, Zaria, Nigeria

Iyiola-Tunji AO, Annatte I, Adesina MA, Ojo OA, Buba W, Nuhu S, Bello M, Saleh I, Yusuf AM, Tukur AM, Hussaini AT, Aguiri AO (2015) Evaluation of crop-livestock integration systems among rural farm families at adopted villages of National Agricultural Extension and Research Liaison Services. J Agric Ext 19(2):46–58

Iyiola-Tunji AO, Adesina MA, Ojo OA, Buba W, Saleh I, Yusuf AM, Tukur AM, Bello M, Nuhu S (2017) Characterization of labour usage, harvest and processing activities among crop and livestock farming households at NAERLS-adopted villages. Nig J Agric Ext 18 (1):51–70

Jäkel D (2004) Observations on the dynamics and causes of sand and dust storms in arid regions, reported from North Africa and China. Erde 135:341–367

Jose S (2009) Agroforestry for ecosystem services and environmental benefits: an overview. Agrofor Syst 76:1–10

Kataria RP (2015) Use of feed additives for reducing greenhouse gas emissions from dairy farms. Microbiol Res 6(6120):19–25

Kebede D (2016) Impact of climate change on livestock productive and reproductive performance. Livest Res Rural Dev 28(2). http://www.lrrd.org/lrrd28/12/kebe28227.htm

Kebede A, Tamiru Y, Haile G (2018) Review on impact of climate change on animal production and expansion of animal diseases. Scholars J Agric Vet Serv 5(4):205–215

Khalifa HH (2003) Bioclimatology and adaptation of farm animals in a changing climate. In: Interactions between climate and animal production, European federation of animal science technical series, vol 7. Wageningen Academic Publishers, Wageningen, pp 15–29

Klop G (2016) Low emission feed – using feed additives to decrease methane production in dairy cows. PhD thesis, Wageningen University, Wageningen, NL. 168 pages. ISBN 978-94-6257-894-4. https://doi.org/10.18174/387944

Kurukulasuriya P, Rosenthal S (2003) Climate change and agriculture: a review of impacts and adaptations. Climate Change series paper no. 91, World Bank, Washington DC

Lamidi OS (2005) The use of some non-conventional protein sources for fattening cattle. Unpublished PhD thesis submitted to the School of Post Graduate Studies, Ahmadu Bello University, Zaria

Lawal-Adebowale OA (2012) Dynamics of ruminant livestock management in the context of the Nigerian Agricultural System. Livestock production chapter 4. Intech. pp 62–80. https://doi.org/10.5772/52923

Lindgren J (2019) Reducing Greenhouse gas emissions from cattle production. Friday, October 4, 2019. https://water.unl.edu/article/animal-manure-management/reducing-greenhouse-gas-emissions-cattle-production

McClanahan TR, Cinner JE, Maina J, Graham NAJ, Daw TM, Stead SM, Wamukota A, Brown K, Ateweberhan M, Venus V, Polunin NVC (2008) Conservation action in a changing climate. J Soc Conserv Biol 1(2):53–59

Murphy SP, Allen LH (2003) Nutritional importance of animal source foods. J Nutr 133 (11):3932S–3935S

Myhre G, Samset BH, Schulz M, Balkanski Y, Bauer S, Berntsen TK, Bian H, Bellouin N, Chin M, Diehl T, Easter RC, Feichter J, Ghan SJ, Hauglustaine D, Iversen T, Kinne S, Kirkevåg A, Lamarque J-F, Lin G, Liu X, Lund MT, Luo G, Ma X, van Noije T, Penner JE, Rasch PJ, Ruiz A, Seland Ø, Skeie RB, Stier P, Takemura T, Tsigaridis K, Wang P, Wang Z, Xu L, Yu H, Yu F, Yoon J-H, Zhang K, Zhang H, Zhou C (2013) Radiative forcing of the direct aerosol effect from AeroCom Phase II simulations. Atmos Chem Phys 13:1853–1877

NAERLS (2019) https://naerls.gov.ng/wp-content/uploads/2020/03/Agricultural-Performance-Survey-of-2019-Wet-Season-in-Nigeria-Executive-Summary.pdf

Nardone A, Ronchi B, Lacetera N, Ranieri MS, Bernabucci U (2010) Effect of climate change on animal production and sustainability of livestock system. Livest Sci 130:57–69

Nelson KS, Bwala DA, Nuhu EJ (2015) The Dromdry camel: a review on the aspects of history, physical description, adaptations, behaviour/lifecycle, diet, reproduction, uses, genetics and diseases. Niger Vet J 36(4):1299–1317

Nicholson SE, Palao IM (1993) A re-evaluation of rainfall variability in the Sahel. Part I. Characteristics of rainfall fluctuations. Int J Climatol 13(4):371–389

Nwosu CC, Ogbu CC (2011) Climate change and livestock production in Nigeria: issues and concerns. Agro Sci J Trop Agric Food Environ Ext 10(1):41–60

Ojo O (1977) The climate of West Africa. Heineman, London

Olagunju TE (2015) Drought, desertification and the Nigerian environment: a review. J Ecol Nat Environ 7(7):196–209

Onah NG, Alphonsus NA, Ekenedilichukwu E (2016) Mitigating climate change in Nigeria: African traditional religious values in focus. Mediterr J Soc Sci 7(6):299–308. https://doi.org/10.5901/mjss.2016.v7n6p299

Oseni S, Bebe O (2008) Climate change, genetics of adaptation and livestock production in low-input systems. Paper presented at ICID+18 2nd international conference: climate, sustainability and development in semi-arid regions August 16–20, 2010, Fortaleza–Ceará, Brazil and Hoffmann I (2008) Livestock genetic diversity and climate change adaptation. Livestock and global change conference proceeding. May 2008, Tunisia

Riahi K, Rao S, Krey V, Cho C, Chirkov V, Fisher G, Kindermann G, Nakicenovic N, Rafaj P (2011) RCP 8.5 – a scenario of comparatively high greenhouse gas emissions. Climate Change 109:33–57

Rojas-Downing MM, Nejadhashemi AP, Harrigan T, Woznicki SA (2017) Climate change and livestock: impacts, adaptation and mitigation. Clim Risk Manag 16:133–144

Rowlinson P, Steele M, Nefzaoui A (2008) Livestock and global climate change: adaptation I and II. In: Rowlinson P, Steel M, Nefzaoui A (eds) Livestock and global climate change conference proceeding. Cambridge University Press, Tunisia, pp 56–85

Schmidt MI (2008) The relationship between cattle and savings: a cattle-owner perspective. Dev South Afr 9(4):433–444

Schwanghart W, Schutt B (2008) Meterological causes of Harmattan dust in West Africa. Geomorphology 95:412–428

Smith JP, Pearce BD, Wolfe MS (2012) Reconciling productivity with protection of the environment: is temperate agroforestry the answer? Renewable Agric Food Syst 28(1):80–92

Srivastava NSL (2006) Farm power sources, their availability and future requirement to sustain agricultural production status of farm mechanization in India. IASRI report, 57-68

Suleiman M (2017) Sahel region, Africa. The Conversation Newsletter. https://theconversation.com/sahel-region-africa-72569

Thomson AM, Calvin KV, Smith SJ, Kyle GP, Volke A, Patel P, Delgado-Arias S, Bond-Lamberty B, Wise MA, Clarke LE, Edmonds JA (2011) RCP4.5: a pathway for stabilization of radiative forcing by 2100. Clim Chang 109(1–2):77–94. https://doi.org/10.1007/s10584-011-0151-4

Umar MA, Dalhatu M, Bello A, Nawawi H (2013) Animal traction as source of farm power in rural areas of Sokoto State, Nigeria. Health Safety Environ 1(1):23–28

Umunna MO, Olafadehan OA, Arowona A (2014) Small ruminant production and management systems in Urban area of Southern Guinea Savannah of Nigeria. Asian J Agric Food Sci 2(2):107–114

van de Haar MJ, St. Pierre N (2006) Major advances in nutrition: relevance to the sustainability of the dairy industry. J Dairy Sci 89(4):1280–1291

van Vuuren DP, Edmonds J, Kainuma M et al (2011) The representative concentration pathways: an overview. Clim Chang 109:5–31. https://doi.org/10.1007/s10584-011-0148-z

van Zijderveld SM, Fonken B, Dijkstra J, Gerrits WJJ, Perdok HB, Fokkink W, Newbold JR (2010) Effects of a combination of feed additives on methane production, diet digestibility, and animal performance in lactating dairy cows. J Dairy Sci 94:1445–1454. https://doi.org/10.3168/jds.2010-3635

Vince G (2012) Sucking CO_2 from the skies with artificial trees. Future. BBC Program 4th October, 2012. https://www.bbc.com/future/article/20121004-fake-trees-to-clean-the-skies

7

Agroecology and Climate Change Adaptation: Farmers' Experiences in the South African Lowveld

Cryton Zazu and Anri Manderson

Contents

Abstract

Motivated by interest to increase the resilience of smallholder farmers to adapt to climate change through uptake of agroecology, two community development organizations commissioned a project evaluation upon which this book chapter is written. The chapter discusses how smallholder farmers were experiencing

---

C. Zazu (✉)
Environmental Learning Research Centre, Rhodes University, Grahamstown, South Africa
e-mail: anri@hoedspruithub.com

A. Manderson
Hoedspruit Hub, Hoedspruit, South Africa

implementing agroecology, trying to understand the reasons for adopting such an approach to farming. The chapter also explores and problematizes the relationship between trends in adoption of agroecology and the smallholder farmers' awareness of climate change and adaptation. The chapter confirms that agronomic and income generation are the key reasons for adoption of agroecology. Most of the farmers reminisced about how their crop yields had declined and soils no longer producing enough to feed the family. Other motivating factors for uptake of agroecology included lack of employment, limited income sources, access to health organic foods, and medicinal value of herbs grown. The chapter further concluded that the correlation between adoption of agroecology and farmers' awareness of it as a climate change adaptation measure is generally weak. Smallholder farmers adopted agroecology more for responding to issues of food security, than any conscious desire to adapt to climate change. Implications of this observation is that practitioners working with smallholder farmers need to rethink their approaches and design of interventions to integrate climate change education and learning, so that strong connections between the agroecological practices promoted and adaptation to climate change are made. Such an approach has potential to improve the sustainability and value of the agroecological practices adopted.

Keywords

Agroecology · Permaculture · Climate change · Smallholder farmer · Adaptation · Mitigation

Introduction

Social development organizations have noticed an increase in agroecological activity among smallholder farmers in the South African lowveld in the last 2 years, including in the districts of the Kruger to Canyons Biosphere Region. Agroecology has also become the point of convergence for a network of institutions based in the biosphere and focusing on a range of environmental, climate change, and social concerns. These institutions included the Hoedspruit Hub, Association for Water and Rural Development (AWARD) Mahlatini, Ukuvuna, Kruger to Canyons Biosphere, CHoiCE Trust, Hlokomela, SANParks, and others, who have mandates ranging from agricultural training, poverty alleviation, biodiversity, and water conservation through to health, yet for different reasons have all found agroecology a critical component to realizing these mandates and notably building farmers resilience to adapt to climate change. It is against this context that two of these organizations working in partnership to promote agroecology among smallholder farmers in this area, collaborated to develop this chapter with the aim of sharing the findings of an evaluation conducted to understand the participating smallholder farmers' experiences of agroecology as both a livelihood option and a strategy for climate change adaptation.

Background Information

Kruger to Canyons Biosphere Region encompasses parts of the Limpopo and Mpumalanga provinces, as well as three southern African biomes: grasslands, Afro-montane forests, and the savannah of the lowveld. And as shown in Fig. 1 below, the region borders with the vast Greater Kruger National Park, which is home to a diversity of flora and fauna. The Kruger to Canyons region also include much of both the upper and lower sub-catchment areas of the Olifants river. Major land-use practices in this region include conservation nature reserves areas, mining (gold, phosphate, copper), exotic plantations, and the extensive nonorganic cultivation of subtropical fruits and vegetables (mainly for export) and to a lesser extent peasant farming.

Perhaps the most important contextual history of this area, when considering the development of smallholder agriculture in post-apartheid South Africa, is the remnants of the former apartheid homelands or bantustans. These were areas the apartheid government set aside for African indigenous people to live after they were forcibly removed from urban areas. It was thus the mechanism with which the government realized segregation, but with which they also successfully created labor reservoirs for the mines and other South Africa industries active at the time. It is essential to take into consideration that although the apartheid government

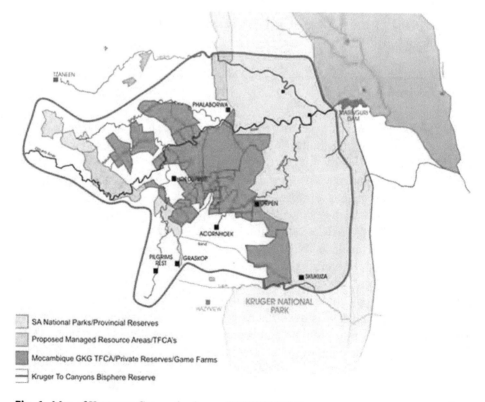

SA National Parks/Provincial Reserves

Proposed Managed Resource Areas/TFCA's

Mocambique GKG TFCA/Private Reserves/Game Farms

Kruger To Canyons Bisphere Reserve

Fig. 1 Map of Kruger to Canyon landscape (AWARD 2016)

intended for these areas to eventually become independent, they were not developed and relied entirely on the larger South African economy.

Vibert (2018) wrote a poignant piece about the effects of forceful removals during apartheid, which displaced people and gathered them in the former homelands. She writes specifically about one homeland that was partly located within the current biosphere, named Gazankulu. It "was envisioned as a rural enclave for women, children, and elderly people of Tsonga ethnicity – their men were the labour force in the mines and cities. This rural-urban binary is misleading: people and resources circulated among these spaces, within and in infraction of the strict spatial regulations of apartheid. Yet rural space was, under apartheid, a space apart" (Vibert 2018). Vibert writes how women recall arriving in this new space where very little preparations had been made for their arrival, finding it inhospitable. They built houses during the day and cooked at night to start a new life. They did not have enough allocated space to grow crops such as sorghum, which they had grown before.

> ...We can't forget sorghum,' the indigenous grain they no longer have space or labour to grow. Mamayila says it's 'very painful [va va ngopfu] to remember the way we were situated. It was so nice. You had enough land to have your garden, donkeys, cattle kraal, one side for goats, one side for pigs.' Today we have 'maybe a cattle-kraal size' says Sara. (Vibert 2018)

Since the end of apartheid, not enough has been done to address the social complexities and trauma of the forceful removals or the underdevelopment of these former homelands, resulting among other social challenges, in millions of unemployed people. Given the legacy of Apartheid's forced removals, villages in Gazankulu are densely populated (Wright et al. 2013). And overutilization of natural resources combined with a lack of proper management of these resources has led to soil erosion and a loss of soil moisture and soil nutrients. These factors, together with low rainfall and poor soils in some areas, have affected smallholder farmer's capacity to produce enough food (AWARD 2016) and with climate change, their vulnerabilities are likely to increase.

Therefore, the work being done by the two development organizations is implemented within this context where historical segregation and limited formal development have left millions of people heavily dependent of agriculture and social grants as livelihoods options.

The Partnership

One of the two development organization is registered as a private company but operates as a social enterprise and training center in the Hoedspruit region. Its social development activities include a high school bursary program, and the training on agroecology, which is of interest to this chapter. The course is meant to equip smallholder farmers with skills and knowledge needed to practice agroecology as a form entrepreneurship along the organic food value chain.

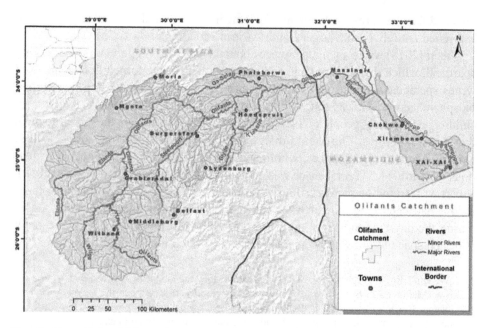

Fig. 2 Map of the Olifants river sub-catchment in which agroecology projects are being implemented (AWARD 2016)

The other organization is a nonprofit organization specializing in multi-disciplinary, participatory, research-based project implementation aimed at addressing issues of environmental sustainability, inequity, and poverty, in particular relating to water conservation and management in the face of climate change. This organization's main geographical area of focus, although not exclusively limited to, is in the areas lying within the lower and upper catchment of the Olifants river – a major tributary of the Limpopo river, which is an international watercourse shared between South Africa and Mozambique. Figure 2 below is a map showing the entire Olifants river catchment.

Changes in Climate

Also, of interest in this chapter is how climate change is being experienced in southern Africa and how this will impact on agriculture. A recent piece in the Farmer's Weekly by Lindie Botha, drawing on the opinion of Prof. Francois Engelbrecht, chief researcher for climate studies, modelling, and environmental health at the Council for Scientific and Industrial Research (CSIR), states that "the average temperature increase in southern Africa due to climate change, is taking place at twice the global rate. The resulting lower rainfall figures and increase in the number of heatwaves will see agricultural landscapes shifting and veld fires growing in frequency. All of this will demand careful planning" and a type of agriculture that is climate smart (Botha 2019). She writes that the Department of Science and

Technology launched the South African Risk and Vulnerability Atlas (SARVA) in response to the changing climates observed in the last few decades. SARVA published in 2018 that rural areas in South Africa are particularly vulnerable to climate change due to its dependence on water and agriculture.

It is against this background that the two partner organizations have, over the course of the partnership, trained and supported a total of 300 smallholder farmers to implement agroecology. And the need to learn more about how the target smallholder farmers are experiencing their implementation of agroecology is what motivate the evaluation, whose findings are discussed in this chapter

Questions Investigated

In order to gain an in-depth understanding of the farmers' experiences of adopting and implementing agroecology, the project evaluation against which this chapter is developed explored the following questions:

1. Why do smallholder farmers adopt an agroecological approach to farming?
2. Is adoption of agroecology as strategy for adaptation linked to farmers' awareness of climate change itself?

Exploring these evaluative questions was done in order to confirm if the agroecological approach is indeed the most effective way to not only meet organizational goals but to enhance the building of smallholder farmers' resilience to adapt to climate change. The answers to the questions of this evaluation are thus the basis for discussions presented in this chapter, the ultimate aim being to improve the way community development organizations work with and support smallholder farmers in South Africa and beyond.

The evaluation was conducted using a qualitative interpretive research approach. The decision to use a qualitative methodology was in sync with the nature of the evaluative questions that the two organizations sought to answer and also the ontological world views of what constitute reality held by the researchers (Creswell 2009; Niewenhuis 2007). Ontology, as defined by Niewenhuis (2007), refers to how one perceives reality or think of that which can be known. Similarly, Patton (2002, 2014) argued that ontology is concerned about the constitution of reality, in the case of this study, "farmers views and experiences of agroecology," and what we are able to know about it. Creswell (2009) also pointed out that ontological world views often shape the orientation and design of the evaluation methodology preferred. Hence the qualitative interpretive paradigm within which this evaluation was conducted, and the case study method used together with the data collection and analysis techniques all reflects the ontological viewpoint of knowledge as socially constructed (Denzin and Lincoln 2011; Maxwell 2010; Yin 2013). An evaluation methodology that does not only focus on the numbers of farmers trained and now then practicing agroecology was needed. Therefore, knowledge interest that required much more than just statistics for donor reporting determined the use of qualitative methodologies. And a qualitative approach allowed the evaluators to generate more insights into farmers experiences illuminating light on some of the grey areas where future programming needed to resolve in order to achieve not only more but sustainable impact, working with smallholder farmers in Limpopo to adapt to climate change.

Using the Case Study Evaluation Method

According to Harrison et al. (2017), case study research has grown in reputation as an effective means to explore and understand complex issues in real-world settings. It has been widely used across several disciplines, particularly the social sciences, education, business, law, and health, to address a wide range of research questions.

The case study method used in this evaluation allowed for a closer examination and analysis of each participating smallholder agroecology farmer in real-life contexts (Harrison et al. 2017). It also allowed the evaluators an opportunity to select information rich participants or cases (Flyvbjerg 2006, 2011) making it possible to generate enough data and in-depth insights into how agroecology is being experienced by and changing the lives of smallholder farmers. It also allowed them to probe in detail the extent to which farmers understand the connection between adopting agroecology and building their resilience to climate change.

Fifteen carefully selected cases of smallholder farmer households drawn from 300 farmers trained and being supported to implement agroecology. These households were purposively selected on the basis of having received training and actively implementing agroecological practices. Purposive sampling meant that the researchers had to pick on the most productive sample (smallholder farmers) that could provide adequate data to answer the research questions (Marshall 1996; Flyvbjerg 2006, 2011). It also allowed for the selection of what Yin (2014) referred to as *data rich* participants. And because this was not a comparative analytical evaluation, selecting those data-rich cases made much epistemological sense.

As represented in Table 1, the farmers involved in the evaluation included those that had been trained through 17 Shaft Training in 2015 and 2016 and those trained by Hoedspruit Hub in 2016 and 2017.

Evaluation information was collected using semi-structured interviews integrated with narrative enquiry and field observation. Yin (2014) argued that semi-structured interviews allow a researcher greater freedom to pursue unexpected, but interesting and relevant comments to a greater depth. Instead of a scripted list of questions, the researcher has a good idea of the questions she would like to ask, and perhaps even an interview guide, but may ask the questions in a way that fit the context of the emerging conversation between researcher and participant smallholder farmer (Yin 2014; Creswell 2009). Also important for a qualitative evaluation such as this one the use of semi-structured interviews allowed the farmers to answer questions in their own words, potentially adding details to the data that might have been missed otherwise (Yin 2011). The interview guide used to collect demographic information relating on participants profiles also provided a way of generating quantitative data

Table 1 The smallholder farmers participating in the evaluation and their training

Code name	Training received
1	Leadership in Agroecology
2	17 Shaft Training (2016)
3	Organic Mango Production
4	17 Shaft Training (2015)
5	Organic Mango Production
6	Herb Gardening
7	Leadership in Agroecology
8	Herb Gardening
9	17 Shaft Training (2016)
10	17 Shaft Training (2015)
11	Entrepreneurship in Agroecology (2016)
12	17 Shaft Training (2016)
13	Entrepreneurship in Agroecology (2016) (not interviewed yet)
14	Leadership in Agroecology
15	Agroecology for the Youth

on harvest and income trends being experienced by smallholder farmers implementing agroecology.

Information collection also entailed use of narrative enquiry where farmers were asked to tell their stories of change. Dyson and Genishi (1994) asserts that storytelling provides a useful theoretical lens through which to examine the ways in which individuals experience the world as illustrated through their own personal stories. Narrative inquiry helped the evaluators to capture the farmers' full experiences of their adoption and implementation of agroecological farming practices. Davis (2007) also argued that storytelling is a very useful way of collecting data especially from informants with low literacy levels such as the case with most of the farmers involved in this project evaluation. As such storytelling helped to make this project evaluation socially inclusive in orientation.

With the consent of participants, their stories and responses to interviews were audio recorded for data transcription and analysis. Some of the interviews were video recorded and footage stored for the future development of a short video film to support the sharing of the emerging findings of the project evaluation.

In order to deepen insights and triangulate information collected through the interviews (Shenton 2004), field observations were conducted and entailed visiting homesteads of all the 15 farmers to learn more about their agroecological practices and ascertain harvests being experienced. Photography was used to capture observations made during the field trips.

In total, 15 interviews were done, and visits to all the 15 farmers undertaken. The data generated was processed through transcribing, translation into English, making it ready for analysis

Information collected was subjected to initial analysis using Atlas Ti 7 to both locate and code the data (Friese 2014). Dohan and Sanchez-Jankowski (1998)

argued that coding data with a well-designed computer program like Atlas Ti 7 can be very useful but not an end in itself. Analysis of data was thus continued using thematic data analysis where emerging themes relating to questions being pursued in the evaluation and discussed in this handbook chapter were identified and analyzed.

Evaluation Findings

Farming Sites

The 15 smallholder farmers who participated in this project evaluation were drawn from the villages as shown in the map below. All these villages are adjacent to the Kruger to Canyons Biosphere area and falls within Maruleng, Lepele Nkhumpi, and Elias Motshoaledi municipalities of Limpopo province. In terms of climate, these areas are generally quite dry with an annual rainfall of around 500 mm/year concentrated in 4 months during the summer (AWARD 2016).

Farmers were selected from villages such as Sidawa, Zebedela, Turkey, and Mametja of Maruleng, and Motetema, Tarfelskop, Makweng, Dithabaneng, Makushoaneng, and Monsterlus of Carpricorn and Elias Motshoaledi (Fig. 3).

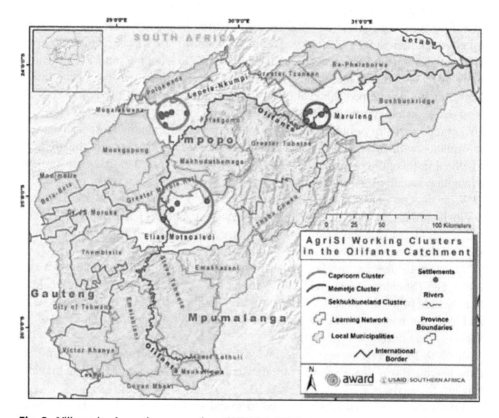

Fig. 3 Villages implementing agroecology (AWARD 2016)

Table 2 Smallholder farmer demographics

Code	Age	Gender	Size of family	Years farming	Production area (ha)
1	48	Female	3	6	0,540
2	41	Female	6	2	0,035
3	53	Female	4	1	0,534
4	44	Female	5	26	0,500
5	56	Male	4	1	0,003
6	40	Male	5	10	0,500
7	52	Female	5	2	0,002
8	58	Female	3	2	0,250
9	20	Female	5	0.2	N.A.
10	59	Female	3	10	N.A.
11	28	Female	2	2	0,200
12	30	Male	5	3	0,420
13	39	Female	3	7	0,200
14	44	Female	6	3	0,500
15	48	Male	2	4	0,032
Averages	45,3		4	5,6	0,28

Demographics

Of the 15 farmers interviewed, 11 were female and 4 were male. Average age of the smallholder farmers interviewed was 45.3, with the youngest being 20 years old. On average, each household size was reported as made up six family members.

The plot sizes for most farmers ranged from 0,002 ha (an area of 4 m × 5 m) being the smallest to the largest of 0,54 ha. Each farmer was producing a good mix of vegetables with limited fruit and herbs. Those farming herbs were benefiting from the market linkage support that Hoedspruit Hub was providing. Table 2 illustrates the above demographics of the farmers.

Reasons for Adopting Agroecological Farming Practices

The chapter confirms that most of the farmers interviewed has various reasons for practicing agroecology. These reasons ranged from the need to produce more food for household consumption, desire to generate income, and influence from neighbors to the farmers' realization of declining yields.

Asked to share his motivation for taking up and practicing agroecology one of the farmers was quoted saying:

> Remember the first thing which is needed by the family before they do any job is food. So for me, I work for food first and then seek money elsewhere. (F2)

Similarly, the other farmer interviewed weighted in by pointing out that:

When I started implementing the permaculture (agroecology) ideas, I noticed that my vegetable production doubled, compared to the first yield where I had little knowledge. This method is cheap and sustainable long term. (F4)

Asked to explain why adopting agroecology and converting from the usual conventional farming, another farmer confidently said that:

I have a diversity of activities in order to produce a wide variety of results such as generating income, providing my family with meat and vegetables. (F7)

Two of the interviewed farmers were despite trained, however, not yet practicing agroecology. The two perceived agroecology as for mainly very small farmers, thus not very suitable for their scale of farming.

It became quite clear that the smallholder farmers interviewed were practicing agroecology because of different reasons. These as reflected above included the both the need to feed their families and earn income from selling the surplus backyard gardening produce. Producing more food to feed the family and income generation as, confirmed by responses from 66% (n = 10) of farmers interviewed, emerged the top two reasons influencing adoption and practicing of agroecology. The desire to improve household income as a reason for practicing agroecology was made much more explicit by a farmer who shared her story as quoted below.

Life with my children and no support from their father was not a good life for me at all. I tried many things to improve the living conditions of my children. I received government benefit but it was not enough to send them to school, provide medical support and to buy the right food. One of the areas I got passionate about was farming, but I had no skills. But then I got training in agroecology and started from a small base growing a few vegetables and crops. . ... and now I earn more money and I am happy. . .. (F10)

The desire to produce food for the family, to share with neighbors, and sell the surplus to generate income is illustrated in Table 3.

Other influencing factors included realization that their soils are no longer producing as much as they used to, keeping up with the jones (learning from neighbors) and as well lack of employment opportunities especially among the youth in the lowveld. About 33% (n = 5) of active agroecological farmers were unemployed and many cited this as the original reason for starting their agroecological production units. For many farmers, continued unemployment meant no other option than to continue farming agroecologically. One farmer alluded to the medicinal value of herbs as the reasons for his interest in doing agroecology. Access to market and increasing demand for organic vegetables by local lodges were also mentioned as factors shaping the way agroecology was being adopted and implemented. Also, very interesting is that a few farmers thought that agroecology is easier and cheaper to do as it does not require heavy use of chemicals and fertilizers. Ecological reasons for implementing agroecological farming practices

Table 3 Household consumption, sharing, and selling of agro-produce

Code	Household use (%)	Sharing (%)	Selling (%)	Income p.m. (rands)
F1	54	22	24	900
F2	20	80	0	0
F3	20	0	80	600
F4	50	0	50	5000
F5	10	10	80	1200
F6	20	0	80	800
F7	20	40	40	200
F8	10	0	90	300
F9	N.A.	N.A.	N.A.	0
F10	N.A.	N.A.	N.A.	0
F11	100	0	0	0
F12	60	0	40	420
F13	40	5	55	5100
F14	90	5	5	120
F15	20	10	70	600
Avg	33,77	16,88	49,33	1016,00

were also observed and some of the farmers talked of the need to improve soil health and save water. Asked to explain why she adopted agroecology, one farmer felt agroecology enabled communities to live in harmony with nature, and she said that,

> To me it would be beneficial if more people can farm agroecologically and if they can impart it to their children. People say they are going to pray for rain. We are not living harmoniously with nature, and not sending rain is nature's response. It's how nature talks

The desire to save water which was also linked to ecological reasons for taking up agroecology could be easily understood from the fact that most of the farmers in the area covered by this evaluation (lowveld of Limpopo) did not have reliable water supply and were actually buying water (Award 2016) to sustain their gardens. Water was therefore even without or before onset of climate change a scarce resource. Others talked of social motivations including that gardening kept them physically active and healthy, reducing stress, and keeping them out of poverty and criminal activities.

The observations made above were also reported by similar research studies. Studies done by Nilesa and Mueller (2016) revealed that farmers change their agricultural behaviors not only because of the changes in climatic conditions. The evaluation findings revealed that factors such as the inherent desire to sustain food security and the changes in the wider environment, e.g., changes in soil fertility and productivity levels, contributed to the changes in agricultural practices adopted by farmers. Scholars such as Toffolini et al. (2018) also revealed that the evolution of farmers' practices towards agroecology is mainly influenced by agronomist factors to increase food production. In a similar sense, Hubert (2012) also observed that agroecology has, in various countries, been considered a viable option for achieving sustainable food production systems. Accordingly, Altieri (1999) claimed that the agroecological approach provides an alternate path to increasing crop production because of its reliance on local farming knowledge and technologies suited to different and marginal climatic conditions such as those of Limpopo.

The Connection Between Agroecology and Climate Change

Out of the 15 farmers interviewed in this project review, only one (n = 1) of them made an explicit reference to climate change. This is a very interesting observation but one that is unusual. Such an observation can be understood from what other studies has revealed. Nilesa and Mueller (2016) pointed out that they are unaware of any studies that have examined the extent to which farmers' perception of climate change directly explains their changes in farming practices. Whyte (2014) also observed that even though the broad issue of adaptation to environmental change is not new for many indigenous peoples, it is something done out of survival instincts rather than any significant levels of climate change awareness. Closely related, Mugambiwa (2018) concluded that subsistence farmers have always adopted adaptive strategies to changes such declining crop yields. In his study, he concluded that changes in farming practices such as a shift from maize to traditional millet and sorghum that farmers in Mutoko, a district in Zimbabwe, adopted were triggered by the desire to preserve local indigenous knowledges and cultures rather than the changing climatic conditions. However, it can be argued that this change and adoption of indigenous knowledge practices as acknowledged by Waha et al. (2013) to respond to frequent droughts, scarcity of rain, and decreased crop yields is by default a strategy for adapting to climate change.

Other studies done by Arbuckle et al. (2013) and Niles et al. (2015) to investigate farmers' perceptions of climate change, its risks, and potential to influence adoption of adaptation and mitigation behaviors also confirmed that farmers relate more to weather than climate change. This observation is also reflected in what one farmer who made reference to climate change was quoted saying,

> Then I ask myself, isn't it important to know, what is hindering rain from coming? So, all these practices we are doing, it's hindering rain from coming, by contributing to climate change. We can pray and pray and pray, but if we don't change our thinking, decolonise our minds, rain will not come. We can expect drought

Farmers are therefore arguably more affected by weather-related losses than climate change per se. Thangata et al. (2002) argued that the decisions made by smallholder farmers to adopt agroecological practices, e.g., agroforestry, in the context of Zambia reflected the farmers' perceptions of worst-case weather changes such as delayed rainfall or droughts. And thus, if they were asked about how weather determines their farming practices, a good number of the farmers involved in this evaluation could have provided more explicit answers. This finding denotes the need to appreciate that for most farmers the adoption of agroecology is, at experience level, motivated more by other factors than solely climate change (Palm et al. 2010; Beddington et al. 2011; Nyanga et al. 2011).

While agroecology has been widely associated with strengthening the resilience of farmers and rural communities (Niles et al. 2015), evidence to support that adoption of agroecological practices is a direct result of the farmer's awareness of climate change remains anecdotal. Nyanga et al. (2011), in a study done in Zambia, also reported of a positive correlation between perception of increased droughts and adoption of conservation agricultural practices such as agroecology but no correlation between farmer' attitudes towards climate change itself. The increased interest in agroecology has also been not only linked to climate change but other narratives such as the green revolution. Hence the tendency to give function to agroecology and all its associated soil, water, and biodiversity conservation technologies, as a strategy for climate change adaptation, can only be referred to as of zero interest from the perspective of farmers. Other scholars have further critiqued the way agroecology has now been hijacked by the politics of the day and repackaged as climate smart agriculture. The implication of the observation made here is that it is just as important to know the priorities of the farmer implementing agroecology practice than to present it as solely a climate change adaptation measure.

Conclusion

While the work done by the two development organizations who commissioned the writing of this chapter is aimed at building resilience for climate change adaptation, it is important to note that farmers' decisions to adopt agroecology is influenced by many other factors. This observation gives added value to the need for additional interventions such as the dialogues for climate change literacy and adaptation (DICLAD) which can improve farmers' levels of knowledge and awareness of climate change, and how this relates to adopted agroecological practices being implemented.

In general, the findings discussed in this chapter also calls for a redesign of contemporary farmer extension support services. There is need to adopt approaches that not only empowers farmers with basic knowledge of climate change science but also recognize that agroecological practices and technologies have a history that is tied to a range of evolving farmer's priorities. Working with farmers and getting them to a level of consciousness that recognizes the value of their indigenous agroecological knowledges and systems for more than just climate change adaptation is what must be pursued. Therefore, advancing the climate change adaptation dimension of agroecology at the expense of the other equally important roles and values is not only reductionist but problematic.

References

Altieri MA (1995) Agroecology: the science of sustainable agriculture. Westview Press, Boulder
Altieri MA (1999) Applying agroecology to enhance the productivity of peasant farming systems in Latin America. Environ Dev Sustain 1:197–217
Arbuckle et al (2013) Climate change beliefs, concerns, and attitudes toward adaptation and mitigation among farmers in the Midwestern United States. Clim Change 117:943–950
AWARD (2016) Grain-SA smallholder farmer innovation Programme, annual report (2015–2016). Limpopo, South Africa
Beddington et al (2011) Achieving food security in the face of climate change: summary for policy makers from the commission on sustainable agriculture and climate change. In CGIAR Research Program on Climate Change. Agriculture and Food Security (CCAFS), Copenhagen. Available on http://www.ccafs.cgiar.org/commission. Accessed 10 June 2020
Botha L (2019) Climate change: it's happening faster than you think. Farmers Weekly, 7 January. Available: https://www.farmersweekly.co.za/agri-technology/farming-for-tomorrow/climate-change-happening-faster-think/. Accessed 4 Jan 2020
Creswell JW (2009) Research design. Qualitative, quantitative and mixed methods approaches, 3rd edn. SAGE, London
Creswell JW (2013) Qualitative inquiry and research design: choosing among five approaches. SAGE, Thousand Oaks
Creswell JW, Plano CV (2007) Designing and conducting mixed methods research. SAGE, Thousand Oaks
Davies MB (2007) Doing a successful research project, using qualitative or quantitative methods. Palgrave Macmillan, New York
Davis P (2007) Storytelling as a democratic approach to data collection: interviewing children about reading. Educ Res 49(2):169–184
Denzin NK, Lincoln YS (2011) Introduction: the discipline and practice of qualitative research. In: Denzin NK, Lincoln YS (eds) The Sage handbook of qualitative research, 4th edn. SAGE, Thousand Oaks, pp 1–20
Dohan M, Sanchez-J M (1998) Using computers to analyse ethnographic field data: theory and practical considerations. Annu Rev Sociol 24:477–498
Dyson AH, Genishi C (1994) The Need for Story: Cultural Diversity in Classroom and Community. National Council of Teachers of English, 1111 W. Kenyon Rd., Urbana, IL 61801-1096
Flyvbjerg B (2006) Qualitative inquiry: five misunderstandings about case-study research. SAGE, London
Flyvbjerg B (2011) Case study. In: Denzin NK, Lincoln YS (eds) The Sage handbook of qualitative research, 4th edn. SAGE, Thousand Oaks, pp 301–316
Friese S (2014) Qualitative data analysis with ATLAS.ti, 2nd edn. SAGE, London
Gillham B (2005) Cases study research method. Continuum, New York
Harrison H, Birks M, Franklin R, Mills J (2017) Case study research: foundations and methodological orientations. Forum Qual Soc Res 18(1):Art 19. http://nbn-resolving.de/urn:nbn:de:0114-fqs1701195. Accessed 19 May 2020
Hubert C (2012) Agrarian dynamics and population growth in Burundi: Agroecology before its time. core.ac.uk/display/87834265. Accessed 10 June 2019
Marshall MN (1996) Sampling for Qualitative Research. Family Practice 13:522–525. https://doi.org/10.1093/fampra/13.6.522. Accessed 14 Aug 2019
Maxwell JA (2010) Qualitative Research Design: An Interactive Approach (Applied Social Research Methods) 3rd Edition. Sage
Mugambiwa SS (2018) Adaptation measures to sustain indigenous practices and the use of indigenous knowledge systems to adapt to climate change in Mutoko rural district of Zimbabwe. Unpublished PhD study
Muhammad A, Janpeter S, Jürgen S, Farhad Z (2016) Climate change vulnerability, adaptation and risk perceptions at farm level in Punjab, Pakistan. Sci Total Environ 547:447–460

Niewenhuis J (2007) Qualitative research design and data gathering techniques. In: Maree K (ed) First steps in research. Van Schaik Publishers, Pretoria

Niles MT, Lubell M, Brown M (2015) How limiting factors drive agricultural adaptation to climate change agriculture. Ecosystem Environ 200:178–185

Nilesa MT, Mueller ND (2016) Farmer perceptions of climate change: associations with observed temperature and precipitation trends, irrigation, and climate beliefs. Glob Environ Chang 39:133–142

Nyanga P, Johnsen F, Kalinda T (2011) Smallholder 'farmers' perceptions of climate change and conservation agriculture: evidence from Zambia. J Sustain Dev 4(4):73–85

Palm C et al (2010) Identifying potential synergies and trade-offs for meeting food security and climate change objectives in sub-Saharan Africa. Proc Natl Acad Sci 107(46):19661–19666

Patton MQ (2002) Qualitative research and evaluation methods, 3rd edn. SAGE, London

Patton MQ (2014) Qualitative research and evaluation methods. Integrating theory and practice. SAGE, London

Patton MQ (2017) Facilitating evaluation. Principles in practice. SAGE, London

Prokopy et al (2015) Farmers and climate change: a cross-national comparison of beliefs and risk perceptions in high-income countries. Environ Manag 56:492–504

Shenton AK (2004) Strategies for ensuring trustworthiness in qualitative research projects. In Education for Information. London: IOS Press

Terre Blanche M, Kelly K (1999) Interpretive methods. In: Terre Blanche M, Durrheim K (eds) Research in practice. Applied methods for the social sciences. UCT, Cape Town

Thangata P, Hildebrand P, Gladwin C (2002) Modelling agroforestry adoption and household decision making in Malawi. Afr Stud Q 6(1–2):249–268

Thomas DR (2006) A general inductive approach for analyzing qualitative evaluation data. Am J Eval 27(2):237–246. https://doi.org/10.1177/1098214005283748. Accessed 12 May 2020

Toffolini Q et al (2018) Agroecology as farmers' situated ways of acting: a conceptual framework. Agroecol Sustain Food Syst

Vibert E (2018) Healing in the soil. Womensfarm.org. https://www.womensfarm.org/hleketani-garden/healing-in-the-soil/. Accessed 14 Jan 2019

Waha K, Bondeau A, Müller C (2013) Adaptation to climate change through the choice of cropping system and sowing date in sub-Saharan Africa. Glob Environ Chang 23(1):130–143

Whyte KP (2014) A concern about shifting interactions between indigenous and nonindigenous parties in U.S. climate adaptation contexts. Interdiscip Environ Rev, Vol. X, No. Y, XXX

Wright et al (2013) Lower Olifants community health: risks and opportunities. Project report

Yin R (2013) Case Study Research: Design and Methods. SAGE Publications

Yin RK (2009) Case study research: design and methods. SAGE, Thousand Oaks

Yin RK (2011) Qualitative research from start to finish. Guilford Press, New York

Yin RK (2014) Case study research: design and methods. SAGE, Los Angeles

8

Sustainable Food Production Systems for Climate Change Mitigation: Indigenous Rhizobacteria for Potato Bio-Fertilization in Tanzania

Becky Nancy Aloo, Ernest Rashid Mbega and
Billy Amendi Makumba

Contents

B. N. Aloo (✉)
Department of Sustainable Agriculture and Biodiversity Conservation, Nelson Mandela African
Institution of Science and Technology, Arusha, Tanzania

Department of Biological Sciences, University of Eldoret, Eldoret, Kenya
e-mail: aloob@nm-aist.ac.tz

E. R. Mbega
Department of Sustainable Agriculture and Biodiversity Conservation, Nelson Mandela African
Institution of Science and Technology, Arusha, Tanzania
e-mail: ernest.mbega@nm-aist.ac.tz

B. A. Makumba
Department of Biological Sciences, Moi University, Eldoret, Kenya

Abstract

The global rise in human population has led to the intensification of agricultural activities to meet the ever-rising food demand. The potato (*Solanum tuberosum* L.) is a crop with the potential to tackle food security issues in developing countries due to its short growth cycle and high nutrient value. However, its cultivation is heavily dependent on artificial fertilizers for yield maximization which culminates in global warming and other environmental problems. There is need, therefore, for its alternative fertilization technologies to mitigate climate change. This study evaluated the potential of indigenous rhizobacteria for potato cropping in Tanzania. Ten potato rhizobacterial isolates belonging to *Enterobacter, Klebsiella, Citrobacter, Serratia,* and *Enterobacter genera* were obtained from a previous collection from different agro-ecological areas in Tanzania. The isolates were characterized culturally, microscopically, biochemically, and by their carbohydrate utilization patterns. Their *in vitro* plant growth-promoting (PGP) traits such as nitrogen fixation, solubilization of phosphates, potassium, and zinc, and production of siderophores, indole acetic acid, and gibberellic acids were then evaluated. Lastly, sterilized potato seed tubers were bacterized with the inoculants and grown in pots of sterile soil in a screen-house using untreated plants as a control experiment. The potato rhizobacterial isolates had varying characteristics and showed varying *in vitro* PGP activities. The screen-house experiment also showed that the rhizobacterial treatments significantly ($p < 0.05$) enhanced different parameters associated with potato growth by up to 91% and established the potential of most of the isolates as alternative biofertilizers in potato cropping systems in Tanzania.

Keywords

Potato · Biofertilizer · Plant growth promoting rhizobacteria (PGPR) · Climate change · Sustainable agriculture

Introduction

Potato (*Solanum tuberosum* L.) is an important crop for food and economic security in developing countries (FAO 2008). However, its cultivation is heavily dependent on the application of synthetic fertilizers (George and Ed 2011). The general recommendations of synthetic fertilizers for this crop are 120, 123, and 149–199 kg ha^{-1} of nitrogen (N), phosphorus (P), and potassium (K), respectively, for an average fresh tuber yield of 30 t ha^{-1} (Manzira 2011). Nevertheless, these generous fertilizer applications do not always produce the desired results because the crop's rooting system is too shallow for the efficient recovery of fertilizers (Hopkins et al. 2014).

Attempts to establish suitable alternative fertilization mechanisms for crops to minimize environmental impacts and mitigate climate change are quickly gathering momentum worldwide (Kumar et al. 2018). In this context, plant rhizospheres have been the center of attention worldwide for decades. Plant roots secrete nutrient-rich exudates that attract plant growth-promoting rhizobacteria (PGPR) that contribute to plant growth promotion (PGP) directly or indirectly, for instance, through the production of phytohormones and siderophores, solubilization of phosphates, and biological nitrogen fixation (BNF) (Kumar et al. 2018).

Biofertilizers are PGPR-cultures which are developed for use as inoculants to improve soil fertility and plant productivity (Aloo et al. 2019). The PGPR of different crops like legumes have extensively been studied as biofertilizers, but the potato rhizobacterial communities are not yet fully understood, yet they could extremely be important as alternative biofertilizers for this crop. Understanding their interactions with the potato to unravel their PGP potentials for sustainable potato cropping is equally important. It is now known that indigenous PGPRs of specific crops can make better biofertilizers because they are completely adapted and established in specific environments and can be more competitive than introduced inoculants (Sood et al. 2018). Cognizant of this, the present study was designed to explore the PGP functions of indigenous rhizobacterial strains selected from a previous study that had isolated, and identified rhizobacteria from potato rhizospheres and tubers from different agro-ecological regions in Tanzania. The selected rhizobacteria were studied culturally, microscopically, biochemically and based on their carbohydrate (CHO)-utilization patterns. Their *in vitro* PGP functions and effects on various growth parameters of potato in pot experiments were investigated under screen-house conditions, using un-inoculated potato plants as controls. Understanding these bacteria and their functions can enable the identification of suitable inoculants for the development of sustainable potato cropping systems in Tanzania and provide deeper insights into the overall mitigation of climate change in Africa.

Materials and Methods

Rhizobacterial Cultures

Ten rhizobacterial cultures that had previously isolated from potato rhizosphere soils growing in various regions in Tanzania and identified using their 16S rRNA gene sequences (Aloo et al. 2020) were selected for the present study. The strains, sources, and species of these rhizobacterial cultures are displayed in Table 1.

Table 1 Strains, sources, and species of the potato rhizobacterial isolates used in the study

SN	Strain code	Source	Strain identity
1	*LUTS1*	Luteba	*Klebsiella grimontii LUTS1*
2	*LUTS2*	Luteba	*Enterobacter tabaci LUTS2*
3	*MWAKS1*	Mwakaleli	*Klebsiella oxytoca MWAKS1*
4	*MWAKS5*	Mwakaleli	*Enterobacter asburiae MWAKS5*
5	*MATS3*	Matadi	*Enterobacter tabaci MATS3*
6	*MPUS2*	Mpunguti	*Enterobacter tabaci MPUS2*
7	*KIBS5*	Kibuko	*Serratia liquefaciens KIBS5*
8	*MWANS4*	Mwangaza	*Citrobacter freundii MWANS4*
9	*BUMS1*	Bumu	*Enterobacter ludwigii BUMS1*
10	*NGAS9*	Ngarenairobi	*Serratia marcescens NGAS9*

Cultural, Microscopic, and Biochemical Characterization of the Rhizobacterial Cultures

The morphological characterization of the rhizobacterial colonies was performed based on their sizes, shapes, colors, margins, opacity, elevation, and texture as described by Somasegaran and Hoben (1994). The determination of Gram staining properties was performed using the 3% potassium hydroxide (KOH) string test (Pradhan 2016). Rhizobacterial smears were prepared and stained with safranin and observed under oil immersion ($\times 100$) on a fluorescence microscope (Optika B-350) to determine the cell shapes. Evidence of flagella motility was checked in a motility test medium (MTM).

The qualitative assessment for the production of organic acids was determined using the methyl red (MR) (Sambrook and Russell 2001). The catalase test was used to identify isolates that produced catalases and the citrate test was used to detect the ability of the rhizobacterial isolates to utilize citrate as the sole source of carbon and energy on Simmons citrate agar (Simmons 1926). The ability of the isolates to produce hydrogen sulfide (H_2S) was checked on sulfur indole motility (SIM) agar tubes, and the oxidative-fermentative (O-F) test medium was used to evaluate the oxidative and fermentative abilities of the isolates. The production of indole by the isolates was also performed on SIM cultures using Kovac's reagent. Lastly, the CHO utilization patterns of the isolates were glucose, sucrose, mannitol, maltose, dextrose, lactose, fructose, dulcitol, sorbitol, trehalose, cellobiose, and ribose as described by Hugh and Leifson (1953).

In vitro Experiments

The rhizobacterial cultures were screened for various *in vitro* PGP activities. Pikovskaya's medium containing tricalcium phosphate (TCP) (Wahyudi et al. 2011) and Aleksandrov's medium containing potassium alumino-silicate (Sindhu et al. 1999) were used to evaluate for P and K solubilization by the rhizobacterial strains, respectively. Zinc solubilization assays were performed using ZnO as the insoluble Zn source (Fasim et al. 2002). For the quantitative estimation of P, K, and Zn solubilization, the optical densities (OD) of culture supernatants from Pikoskaya's, Alexandrov's, and ZnO broths were determined spectrophotometrically at A_{690}, A_{799}, and A_{399}, respectively, using a multimode reader (Synergy HTX – Biotek). The quantities of solubilized P, K, and Zn in mg L^{-1} were subsequently calculated from standard curves of KH_2PO_4, KCl, and $ZnSO_4$, respectively. The halozones around the bacterial colonies on the plates of respective insoluble compounds were measured and used to compute the solubilization index (SI) for each compound using Eq. 1 (Edi-Premona et al. 1996).

$$\text{Solubilization index (SI)} = \frac{\text{Colony Diameter (cm)} + \text{Halozone Diameter (cm)}}{\text{Colony Diameter (cm)}}$$

$$(1)$$

The production of IAA and GA by the isolates was evaluated as previously described by Vincent (1970) and Holbrook et al. (1961), respectively. The nitrogenase activities of the isolates were checked on solid and liquid N-free media (NFM). The formation of brown or yellow colors in the NFM broth cultures indicated NH_3 production, and its OD was measured spectrophotometrically at 435 nm. The concentration of NH_3 was then estimated by comparing the absorbance of samples with a standard curve of ammonium sulfate in the range of 0.0–10 mg L^{-1} (Goswami et al. 2014). The siderophore production abilities of the isolates were assessed using chrome azurol S (CAS) liquid assays and agar plates as described by Schwyn and Neilands (1987). In the liquid CAS assays, the percent siderophore units (% SU) per isolate were calculated from the absorbance measurements of samples and reference solutions using Eq. 2 (Payne 1993).

$$\%\text{Siderophore units (\%SU)} = \frac{\text{Reference absorbance} - \text{Sample absorbance}}{\text{Reference absorbance}}$$
$$\times 100\% \qquad (2)$$

For the solid CAS assays, each experiment was performed in triplicates and the diameters of the orange or yellow halozones were used to calculate the siderophores production index (SI) using Eq. 3 (Batista 2012).

$$\text{Siderophore production index (SI)} = \frac{\text{Orange halozone (cm)}}{\text{Colony diameter (cm)}} \qquad (3)$$

The Potted Experiment

The rhizobacterial cultures were grown in 50 mL universal bottles filled with 25 mL Tryptic soy broth in a rotary shaker (200 rpm) for 16 h at 28 °C. The absorbance of the bacterial suspensions was evaluated spectrophotometrically at 600 nm using a multimode reader (Synergy HTX – Biotek), and each culture was diluted in sterile distilled water to a final concentration of 1×10^6 CFUs mL^{-1}. The cells were harvested by centrifugation at 4000 rpm for 20 min at 4 °C and resuspended in 100 mL of 7% Carboxy Methyl Cellulose (CMC) solution to help bind the cells to the tubers. Potato seed tubers sourced from a nearby local market were surface-sterilized (using 3% sodium hypochlorite for 3 min and rinsing four times in sterile distilled water). The tubers suspended in the prepared bacterial suspensions for 30 min and sown in plastic pots (20 cm wide) containing 250 g of 24-h oven-sterilized soil with pH: 7.33, electrical conductivity (EC): 207.33 μS cm^{-1}, soluble salts: 0.07%, organic carbon (OC): 0.89%, organic matter (OM): 1.53%, N: 0.08%, zinc (Zn): 35.62 mg kg^{-1}, P: 231.64 mg kg^{-1}, K: 7.59 mg kg^{-1}, iron (Fe): 1.31 mg kg^{-1}, sand: 68.67%, clay + silt: 29.44% and gravel: 1.88%. The experiment was set up in a completely randomized block design with three replicate potato pots per treatment, giving a total of 93 pots for 90 bacterized tubers (10 isolates in three triplicates) and three nonbacterized tubers. The screen house conditions were naturally maintained at 20–22 °C with a day length of 12 h and watered every 48 h using sterile distilled water (150 mL pot^{-1}).

The number of days to emergence (DTE) and flowering (DTF) per treatment was recorded and 90 days after planting (DAP), the crops were harvested and data obtained on the number of tubers, length and weight of shoots, and the average weight and size of tubers per plant. To obtain the average size of tubers per plant, the diameter of each tuber from each potato plant was measured using a measuring tape, and the diameter of tubers per plant obtained by dividing the total diameter of tubers per plant by the number of tubers from that plant. The average radius of tubers per plant was obtained by diving the average diameter by two and used to determine the potato tuber size per plant (Eq. 4).

$$\text{Average tuber size} = \frac{4}{3} \pi \text{ x (Average radius)}3 \qquad (4)$$

The potato rhizospheric soils were evaluated for various physicochemical properties. The pH and EC of the soils were analyzed by the saturated paste method as proposed by Jackson (1973) and Chi and Wang (2010), respectively. The (%) OC in the soils was determined by the potassium dichromate ($K_2Cr_2O_7$) wet digestion method and the (%) OM of the samples was derived from the (%) OC using Eq. 5 (Walkley and Black 1934).

$$(\%) \text{ Organic Matter} = (\%) \text{ Organic Carbon} \times 1.724 \qquad (5)$$

The N content in the rhizospheric soils was determined following the micro-Kjeldahl method (Bremmer and Mulvaney 2015). The Mehlich III method was used to extract the soil P and Zn (Tran and Simard 1993), while the extraction of K was performed using the ammonium acetate method (Jackson 1973). The 1, 10-phenanthroline complex method described by Chaurasia and Gupta (2014) was used for the extraction of Fe in the samples. The quantitative estimation of P, K, Zn, and Fe in the soil extracts was performed by determining the OD spectrophotometrically at A_{690}, A_{799}, A_{399}, and A_{510}, respectively, using the multimode reader (Synergy HTX – Biotek) and subsequently calculating their concentrations from standard curves of KH_2PO_4, KCl, $ZnSO_4$, and ferrous ammonium sulfate, respectively.

The potato tuber nutrient contents were evaluated per plant from single well-developed tubers. The tubers were surface-sterilized using 2% sodium hypochlorite and rinsed four times with sterile distilled water. Next, they were chopped into small pieces using a clean knife and dried in the oven (80 °C) for 5 days. Particle size reduction was performed by mechanically grinding, crushing, and milling them into fine amorphous powders. The P, Fe, and K in them were extracted using 50 mL of 2% acetic acid on 0.2 g of samples, followed by filtering (Miller 1995). The Mehlich III method was used to extract Zn in the powdered potato tuber samples (Tedesco et al. 1995). All sample extracts were subjected to quantification of P, K, Zn, and Fe by absorbance measurements using the multimode reader (Synergy HTX – Biotek), and the respective concentrations were obtained using the standard graphs as previously described for soil analysis. The analysis of total %N in potato samples was performed using the micro-Kjeldahl process (Bremmer and Mulvaney 2015), and their crude protein contents were estimated using Eq. 6 based on the assumption that N constitutes 16% of protein (AOAC 1995).

$$\text{Crude protein content } (\%) = \text{micro} - \text{Kjeldahl N content } (\%) \times 6.25 \quad (6)$$

Statistical Analysis

All statistical analyses were performed using the XLSTAT (Version 2.3, Adinsoft) at a 95% level of confidence. The Shapiro-Wilk test was used to test for normality of data and multiple comparisons of variances were performed using Multivariate Analysis of Variance (MANOVA). Variables with significantly different means were subjected to post hoc analysis using Tukey's Honest Significant Difference (HSD) test. Spearman's correlation was used to evaluate relationships between potato nutrient contents, rhizosphere soil properties, and potato biometrics in the screen house experiment. The percent increase/decrease in levels of different response/dependent variables was calculated from the field experiments using Eq. 7 to assess the differences between treatment and control experiments.

$$\frac{\text{Treatment} - \text{Control}}{\text{Treatment}} \times 100\% \quad (7)$$

Results

The Cultural, Microscopic, Biochemical, and Carbohydrate Utilization Properties of the Potato Rhizobacterial Isolates

The cultural, microscopic, biochemical, and the CHO utilization properties of the potato rhizobacterial isolates are displayed in Table 2. All colonies were round in form except for *E. tabaci* MPUS2 and *S. liquefaciens* KIBS5 which were spreading in form and *C. freundii* MWANS4 which had a rhizoid appearance. They portrayed different colors, textures, and margins and most were opaque. All were rod-shaped

Table 2 Cultural, microscopic, biochemical, and carbohydrate utilization characteristics of the potato rhizobacterial isolates

	LUTS2[a]	MPUS2[b]	MWANS4[c]	NGAS9[d]	KIBS5[e]	LUTS1[f]	MWAKS5[g]	MATS3[h]	MWAKS1[i]	BUMS1[j]
Cultural characteristics										
Form	Round	Spreading	Rhizoid	Round	Spreading	Round	Round	Round	Round	Round
Color	Cream	Cream	Cream	Yellow	Cream	White	Cream	Yellow	Yellow	Cream
Elevation	Raised	Raised	Flat	Raised	Flat	Flat	Raised	Raised	Raised	Raised
Opacity	Opaque	Transparent	Transparent	Opaque	Transparent	Opaque	Opaque	Opaque	Opaque	Opaque
Texture	Smooth	Smooth	Rough	Rough	Smooth	Rough	Rough	Rough	Rough	Smooth
Margins	Regular	Irregular	Undulate	Regular	Irregular	Irregular	Irregular	Regular	Irregular	Regular
Microscopic characteristics										
Shape	Rods	Rods	Rods	Rods	Rods	Rods	Rods	Rods	Rods	Rods
Gram stain	−	−	−	−	−	−	−	−	−	−
Motility	+	+	+	−	+	+	−	+	−	+
Biochemical properties										
MR[k]	+	+	−	+	+	+	−	+	−	−
Catalases	+	+	+	++	++	+	−	+	++	++
Citrate	+	+	+	+	+	+	+	+	−	−
H$_2$S[l]	+	−	−	+	+	+	+	−	−	−
Indole	+	+	+	+	+	+	+	+	+	+
O-F[m]	+	+	+	+	+	+	+	+	+	+
Carbohydrate utilization patterns										
Glucose	+	+	+	+	+	+	+	−	+	+
Sucrose	+	+	+	+	+	+	+	+	+	+
Maltose	+	+	+	+	+	+	+	+	+	+
Fructose	+	+	+	+	+	+	+	−	+	+
Mannitol	+	+	+	+	+	+	+	+	+	+
Lactose	−	−	−	+	+	+	−	−	+	−
Dextrose	+	+	+	+	+	+	+	+	+	+

Ribose	+	+	-	+	+	+	+	+	+
Trehalose	+	+	+	+	+	+	+	+	+
Cellobiose	+	+	-	+	+	+	+	+	-
Ducitol	+	+	+	+	+	-	+	+	+
Sorbitol	+	+	+	+	+	-	+	+	-

[a]Enterobacter tabaci LUTS2
[b]Enterobacter tabaci MPUS2
[c]Citrobacter freundii MWANS4
[d]Serratia marcescens NGAS9
[e]Serratia liquefaciens KIBS5
[f]Klebsiella grimontii LUTS1
[g]Enterobacter asburiae MWAKS5
[h]Enterobacter tabaci MATS3
[i]Klebsiella oxytoca MWAKS1
[j]Enterobacter ludwigii BUMS1
[k]Methyl Red test for organic acids
[l]Test for production of hydrogen sulfide
[m]Oxidative fermentative test

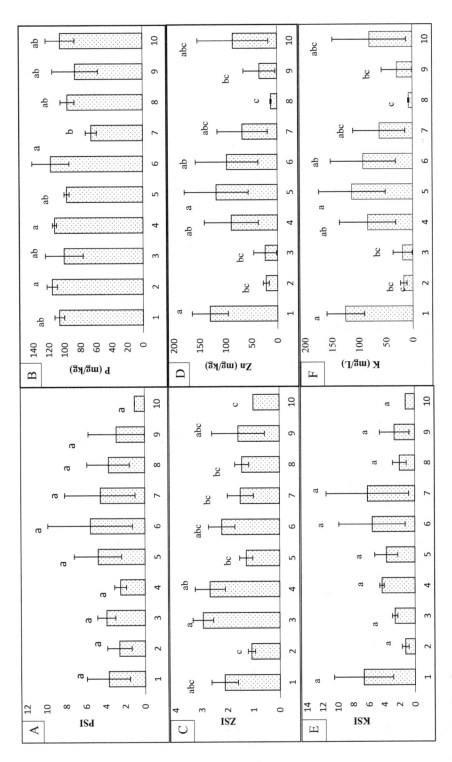

Fig. 1 (continued)

and Gram-negative. Similarly, all strains were motile except for *S. marcescens* NGAS9, *E. asburiae* MWAKS5, and *K. oxytoca* MWAKS1 and all were MR-positive except for *C. freundii* MWANS4, *E. asburiae* MWAKS5, *K. oxytoca* MWAKS1, and *E. ludwigii* BUMS1.

Except for *E. asburiae* MWAKS5, the rest of the studied isolates were catalase-positive, with some isolates like *E. ludwigii* BUMS1, *K. oxytoca* MWAKS1, *S. liquefaciens* KIBS5, and *S. marcescens* NGAS9 exhibiting strong catalase activities than the rest of the isolates. Half of the isolates exhibited H_2S production abilities but indole production and O-F tests were positive for all of them. *Klebsiella oxytoca* MWAKS1, *K. grimontii* LUTS1, and *S. marcescens* NGAS9 metabolized all the tested sugars. *Enterobacter ludwigii* BUMS1, *E. tabaci* MATS3, *E. asburiae* MWAKS5, *C. freundii* MWANS4, *E. tabaci* MPUS2, and *E. tabaci* LUTS2 could not metabolize lactose. Additionally, *E. tabaci* MPUS2 and E. tabaci MATS3 could not metabolize cellobiose, *E. tabaci* LUTS2, *E. tabaci* MPUS2, and *S. liquefaciens* KIBS5 could not metabolize sorbitol and *S. liquefaciens* KIBS5 could not metabolize dulcitol.

In vitro Plant Growth Promoting Abilities of the Potato Rhizobacterial Isolates

The results of *in vitro* solubilization of P, Zn, and K by the 10 potato rhizobacterial isolates are portrayed in Fig. 1a–f. Significant differences among the isolates were observed for ZSI in the qualitative assays and quantities of solubilized P ($p = 0.026$), Zn ($p = 0.031$), and K ($p = 0.031$) in the quantitative assays. However, no significant differences were noted for PSI ($p = 0.885$) and KSI ($p = 0.524$) in the qualitative assays. The averages of PSI, ZSI, and KSI were 3.628 ± 0.420, 1.783 ± 0.764, and 3.619 ± 3.563, respectively, while those for quantities of solubilized P, Zn, and K were 100.33 ± 19.90 mg L^{-1}, 67.897 ± 55.46 mg L^{-1} and 62.897 ± 55.46 mg L^{-1}, respectively. The best P solubilizers were *E. tabaci* LUTS 2, *E. tabaci* MPUS2, and *S. liquefaciens* KIBS5 which recorded average quantities of 115.88, 112.59, and 117.43 mg L^{-1} of solubilized P, respectively. Similarly, *S. marcescens* NGAS9, with average ZSI of 2.94 and *E. ludwigii* BUMS1, with an average of 130.26 mg L^{-1} of solubilized Zn, exhibited the best Zn solubilization abilities in the qualitative and quantitative

←——

Fig. 1 *In vitro* nutrient solubilization abilities of the potato rhizobacterial isolates. On the X axes, 1 = *Enterobacter ludwigii* BUMS1, 2 = *Enterobacter tabaci* LUTS2, 3 = *Serratia marcescens* NGAS9, 4 = *Enterobacter tabaci* MPUS2, 5 = *Citrobacter freundii* MWANS, 6 = *Serratia liquefaciens* KIBS5, 7 = *Klebsiella grimontii* LUS1, 8 = *Enterobacter asburiae* MWAKS5, 9 = *Klebsiella oxytoca* MWAKS1 and 10 = *Enterobacter tabaci* MATS3. (**a**): Phosphorus solubilization index (**b**): Quantity of solubilized Phosphorus (**c**): Zinc solubilization index (**d**): Quantity of solubilized Zinc (**e**): Potassium solubilization index (**f**): Quantity of solubilized potassium. Values are means of three replicates and bars with similar letters are not significantly different (ANOVA + Tukey's HSD; $P < 0.05$)

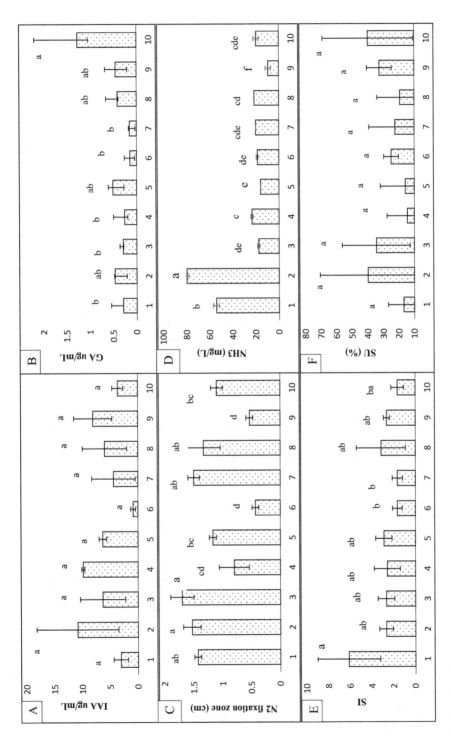

Fig. 2 (continued)

assays, respectively. Only two isolates, *C. freundii* MWANS4 and *E. ludwigii* BUMS1 exhibited good K solubilization abilities in the quantitative assays by yielding an average of 112.98 and 125.26 mg L^{-1} of solubilized K, respectively.

The quantity of IAA produced by the potato isolates *in vitro* was not significantly different ($p = 0.080$) (Fig. 2a). Nevertheless, *E. tabaci* LUTS2 yielded the highest quantity of IAA (10.86 µg mL^{-1}) and the average quantity of IAA produced by all the isolates was 5.57 ± 4.51 µg mL^{-1}. Interestingly, the isolates exhibited significantly different ($p = 0.027$) abilities to produce GA (Fig. 2b). The average quantity of GA produced by the isolates was however only 0.423 ± 0.420 µg mL^{-1} and the best GA producer was *E. tabaci* MATS3 with an average of 1.27 µg mL^{-1}. The isolates exhibited significantly different ($p < 0.0001$) N$_2$-fixation abilities *in vitro* (Fig. 2c, d). The average N$_2$ fixation zones and quantities of ammonia (NH$_3$) recorded for them were 1.153 ± 0.440 cm and 27.97 ± 21.09 mg L^{-1}, respectively. *Serratia marcescens* NGAS9 which recorded an average N$_2$ fixation zone of 1.70 cm and *E. ludwigii* BUMS1 with an average of 79.84 mg L^{-1} produced NH$_3$ yielded the best results from the two tests used to assess for *in vitro* N$_2$ fixation. The isolates exhibited significant differences ($p = 0.021$) with regards to the SI averages (Fig. 2e), but no significant differences ($p = 0584$) were observed with regards to the SU averages (Fig. 2f). The average SI and SU recorded for the isolates were 2.79 ± 1.66 and 26.14 ± 18.25%, respectively. The highest average SI of 6.13 was yielded by *E. ludwigii* BUMS1 and the rest of the isolates yielded significantly lower SI averages ranging from 1.67 to 2.71.

Effects of the Rhizobacterial Treatments on Potato Growth Parameters and Rhizobacterial Soils in the Screen House Experiment

The effects of rhizobacterial treatments on various growth parameters related to potato growth in the screen house experiment are shown in Table 3. Although the DTE averages were not significantly different for the different treatments in this experiment ($p = 0.960$), treatment with some rhizobacteria such as *K. oxytoca* MWAKS1, *E. tabaci* LUTS2, and *E. asburiae* MWAKS5 still resulted in DTE reduction by 7.53%, 3.41%, and 7.53%, respectively. Treatment of the potato seed tubers with the various rhizobacterial isolates significantly ($p = 0.027$) improved the DTF of treated plants in comparison to the untreated controls which recorded the

---

Fig. 2 Effects of rhizobacterial treatments on growth parameters of potted potato plants. On the X axes, 1 = *Enterobacter ludwigii* BUMS1, 2 = *Enterobacter tabaci* LUTS2, 3 = *Serratia marcescens* NGAS9, 4 = *Enterobacter tabaci* MPUS2, 5 = *Citrobacter freundii* MWANS, 6 = *Serratia liquefaciens* KIBS5, 7 = *Klebsiella grimontii* LUS1, 8 = *Enterobacter asburiae* MWAKS5, 9 = *Klebsiella oxytoca* MWAKS1 and 10 = *Enterobacter tabaci* MATS3. (**a**): Quantity of Gibberellic acids (**b**): Quantity of indole-3-acetic acid (**c**): Nitrogen fixation zone (**d**): Quantity of ammonia (**e**): Siderophore production index (**f**): Siderophore production units. Values are means of three replicates and beans with similar letters within the same chart are not significantly different (ANOVA + Tukey's HSD; $P < 0.05$)

Table 3 Effects of rhizobacterial treatments on potato biometric parameters

Treatment	DTE[a]	DTF[b]	Tuber no[c]	Tuber weight (g)[d]	Tuber diameter (cm)[e]	Tuber size (cm^3)[f]	Shoot length[g]	Shoot weight[h]
S1	10.00 (0.00) a	36.67 (−25.44) ab	14.67 (40.90) a	19.93 (91.21) a	3.87 (44.96) a	40.08 (80.20) a	38.00 (53.50) a	13.26 (82.50) a
S2	10.00 (0.00) a	43.33 (−6.16) ab	14.33 (39.50) a	5.39 (68.46) b	2.57 (17.12) bcd	9.13 (43.92) a	37.67 (53.09) a	18.14 (87.10) a
S3	9.333 (−7.53) a	42.00 (−9.52) ab	12.00 (27.75) a	7.40 (77.03) ab	3.53 (39.66) ab	24.67 (79.25) a	36.00 (50.92) a	5.07 (54.24) a
S4	10.00 (0.00) a	36.67 (−25.44) ab	11.67 (25.71) a	4.74 (64.14) b	3.27 (34.86) abc	18.47 (72.28) a	46.33 (61.86) a	6.94 (66.57) a
S5	9.67 (−3.41) a	36.67 (−25.44) ab	12.67 (31.57) a	7.68 (77.86) ab	2.33 (8.58) cd	6.69 (23.47) a	28.33 (37.63) a	19.85 (88.12) a
S6	9.33 (−7.53) a	39.33 (−19.96) ab	12.00 (27.75) a	5.39 (68.46) b	2.87 (25.78) abcd	13.43 (64.27) a	29.00 (39.07) a	8.75 (73.49) a
S7	10.00 (0.00) a	34.67 (−32.68) b	11.33 (23.48) a	7.28 (76.65) ab	2.67 (20.22) bcd	12.83 (60.09) a	28.00 (36.89) a	3.48 (33.33) a
S8	10.00 (0.00) a	38.00 (−21.05) ab	10.00 (13.30) a	2.27 (36.57) b	2.37 (10.13) cd	7.00 (26.86) a	29.00 (39.07) a	3.46 (34.83) a
S9	10.00 (0.00) a	36.67 (−25.44) ab	10.00 (13.30) a	4.30 (60.47) b	2.17 (1.84) d	5.52 (7.25) a	37.67 (36.14) a	3.01 (22.92) a
S10	10.00 (0.00) a	40.67 (−13.11) ab	8.667 (0.00) a	3.87 (56.07) b	2.20 (3.18) d	5.67 (9.70) a	17.00 (−3.94) a	2.19 (−5.94) a

Control[i]	10.667 a	46.000 a	8.667 a	1.703 b	2.133 d	5.117 a	17.667 a	2.323 a
p value	0.960	0.027	0.121	0.010	0.027	0.189	0.169	0.216
Significant	No	Yes	No	Yes	Yes	No	No	No

S1: *Serratia marcescens* NGAS9, S2: *Serratia liquefaciens* KIBS5, S3: *Klebsiella oxytoca* MWAKS1, S4: *Enterobacter tabaci* MATS3, S5: *Enterobacter tabaci* LUTS2, S6: *Enterobacter asburiae* MWAKS5, S7: *Citrobacter freundii* MWANS4, S8: *Enterobacter tabaci* MPUS2, S9: *Klebsiella grimontii* LUTS1, S10: *Enterobacter ludwigii* BUMS1

Values are means of three replicates with (%) increase relative to the control in parenthesis. Means with similar letters within the same columns are not significantly different (ANOVA + Tukey's HSD; $p < 0.05$)

[a] Number of days to emergence

[b] Number of days to flowering

[c] Average number of tubers

[d] Average weight of tubers

[e] Average diameter of tubers

[f] Average size of tubers calculated from the average diameters

[g] Average shoot lengths

[h] Average shoot weights

[i] Un-inoculated control experiment

highest DTF of 46. The rhizobacterial treatments reduced the average DTF of potato plants by approximately 6–33%. The DTF reduction by *C. freundii* of 32.68% was significantly higher than all other rhizobacterial treatments.

Although the number of tubers was not significantly different ($P = 0.121$) across the different rhizobacterial treatments and the un-inoculated control, the rhizobacterial treatments still resulted in crops with increased tuber yields above the control which recorded an average tuber number of 8.67. The greatest average number of tubers of 14.67 was observed from the treatment with *S. marcescens* NGAS9 corresponding to a 91.21% increase relative to the control plants. Significant differences were observed for tuber weight ($p = 0.010$) and diameter ($p = 0.027$) among the different rhizobacterial treatments and the uninoculated control plants.

The highest average tuber weight of 19.93 g corresponding to a 91.21% increase above the control was recorded for potato plants from the treatment with *S. marcescens* NGAS9. This particular treatment also yielded the largest tuber diameter of 3.87 cm corresponding to a 44.96% increase over the control which recorded an average tuber diameter of 2.13 cm.

No significant differences ($p = 0.189$) were observed for tuber sizes for the different treatments and the control experiment. However, treatment with *S. marcescens* NGAS9 produced an average tuber size of 40.08 cm^3 corresponding to an increase of 87.23% above control treatments. The average shoot lengths of potato plants observed for different rhizobacterial treatments in this study were not significantly different ($p = 0.169$). However, all rhizobacterial treatments except for *E. ludwigii* BUMS1 resulted in increased shoot lengths of the potato plants by between 36% and 54% relative to the un-inoculated controls. The treatment with *E. tabaci* MATS3 gave the maximum shoot weight of 46.33 and the highest increment of 61.86% relative to the un-inoculated control. Interestingly, treatment with *E. ludwigii* BUMS1, though not significant, resulted in shoots with lower weights and lengths in comparison to the un-inoculated control. Similarly, the average shoot weights of potato plants observed for different rhizobacterial treatments in this study were not significantly different ($p = 0.126$). However, the treatments still resulted in increased shoot weights of potato plants by up to 82% relative to the control experiments where the average shoot weight was 2.32 g.

The properties of potato rhizospheric soils from the screen house experiment are provided in Table 4. Significant differences ($p < 0.05$) were among the rhizobacterial treatments and the un-inoculated control for all the studied soil properties except for Fe ($p = 0.077$), P ($p = 0.109$), and pH ($p = 0.493$). The treatment with *E. tabaci* MPUS2 resulted in the highest OM content of 3.56% corresponding to a 60% increment over the un-inoculated control. *Klebsiella grimontii* LUTS1 also yielded the best results in terms of EC, salts, and K contents with averages of 1467.33 uS cm^{-1}, 0.51%, and 31.79 mg kg^{-1}, respectively, corresponding to increments of 48.8%, 49.7%, and 75.3%, respectively, above the un-inoculated control. Treatment of potato plants with *E. asburiae* MWAKS5 also yielded significantly higher averages of OM (2.22%), OM (3.83%), and Zn (92.28 mg kg^{-1}) corresponding to increments of 68.3%, 68.4%, and 66.7%, respectively, over the un-inoculated control.

Table 4 Effects of rhizobacterial treatments on physicochemical properties of potato rhizospheric soils

Treatment	OC (%)[a]	OM (%)[b]	pH	EC (μS cm⁻¹)[c]	Salts (%)[d]	N (%)[e]	P (mg kg⁻¹)[f]	Zn (mg kg⁻¹)[g]	K (mg kg⁻¹)[h]	Fe (mg kg⁻¹)[i]
S1	1.91 (63.4) ab	3.29 (63.2) abc	8.04 (−0.4) a	742.00 (−1.2) ab	0.26 (0.0) ab	0.16 (62.5) ab	346.50 (39.3) a	74.20 (56.4) ab	9.58 (16.0) bc	0.86 (98.8) a
S2	1.09 (35.8) ab	1.31 (7.6) d	8.25 (2.2) a	384.33 (−95.4) b	0.14 (−85.7) b	0.07 (14.3) cd	211.47 (0.6) a	54.44 (40.6) ab	14.14 (44.4) bc	1.31 (99.5) a
S3	1.61 (56.5) ab	1.63 (25.8) cd	7.70 (−4.8) a	781.67 (3.9) ab	0.27 (3.0) ab	0.08 (25.0) bcd	282.98 (25.7) a	59.76 (45.9) ab	12.08 (34.9) bc	1.70 (99.4) a
S4	2.04 (65.7) ab	3.51 (65.5) ab	8.12 (0.6) a	459.67 (−63.4) b	0.16 (−62.5) b	0.18 (66.7) a	308.78 (31.9) a	69.23 (53.3) ab	14.71 (46.6) bc	0.86 (99.5) a
S5	2.03 (65.4) ab	3.50 (65.4) ab	8.01 (−0.8) a	771.00 (2.6) ab	0.27 (3.70) ab	0.18 (66.7) a	332.09 (36.7) a	70.41 (54.0) ab	13.07 (39.9) bc	1.489 (99.5) a
S6	2.22 (68.3) a	3.83 (68.4) a	8.11 (0.6) a	581.00 (−29.3) ab	0.20 (−30.0) ab	0.19 (68.4) a	337.99 (37.8) a	97.28 (66.7) a	11.11 (29.3) bc	0.64 (99.3) a
S7	1.98 (64.7) ab	3.42 (64.5) abc	7.70 (−4.8) a	720.67 (−4.2) ab	0.25 (50.0) ab	0.17 (64.7) a	287.83 (27.0) a	77.51 (58.3) ab	13.50 (41.8) bc	0.78 (99.5) a
S8	2.07 (66.2) a	3.56 (66.0) a	5.97 (−35.2) a	755.67 (0.6) ab	0.267 (3.7) ab	0.18 (66.7) a	342.87 (38.6) a	51.88 (37.6) ab	10.09 (22.1) bc	0.89 (99.0) a
S9	1.35 (48.2) ab	1.75 (30.9) bcd	7.48 (−7.9) a	1467.33 (48.8) a	0.51 (49.0) a	0.15 (60.0) abc	331.44 (36.6) a	47.81 (33.3) b	31.79 (75.3) a	0.71 (99.7) a
S10	1.41 (50.4) ab	1.28 (5.5) d	7.91 (−2.02) a	321.33 (−133.7) b	0.11 (−136.4) b	0.06 (0.0) d	356.32 (41.0) a	67.10 (51.8) ab	16.71 (53.0) bc	0.50 (99.6) a

(continued)

Table 4 (continued)

Treatment	OC (%)[a]	OM (%)[b]	pH	EC (μS cm^{-1})[c]	Salts (%)[d]	N (%)[e]	P (mg kg^{-1})[f]	Zn (mg kg^{-1})[g]	K (mg kg^{-1})[h]	Fe (mg kg^{-1})[i]
Control[j]	0.704 b	1.214 d	8.073 a	751.000 ab	B 0.263 ab	0.061 d	21 0.248 a	32.357 b	7.860 c	0.217 a
p value	0.010	0.010	0.493	0.041	0.040	0.007	0.109	0.011	0.001	0.077
Significant	Yes	Yes	No	Yes	Yes	Yes	No	Yes	Yes	No

S1: *Serratia marcescens* NGAS9, S2: *Serratia liquefaciens* KIBS5, S3: *Klebsiella oxytoca* MWAKS1, S4: *Enterobacter tabaci* MWAKS1, S5: *Enterobacter tabaci* LUTS2, S6: *Enterobacter asburiae* MWAKS5, S7: *Citrobacter freundii* MWANS4, S8: *Enterobacter tabaci* MPUS2, S9: *Klebsiella grimontii* LUTS1, S10: *Enterobacter ludwigii* BUMS1

Values are means of three replicates with (%) increase relative to the control in parenthesis. Means with similar letters within the same columns are not significantly different (ANOVA + Tukey's HSD; $p < 0.05$)

[a]Percent organic carbon
[b]Percent organic matter
[c]Electircal conductivity (EC)
[d]Percent salts calculated from EC
[e]Percent nitrogen
[f]Extractable phosphorus
[g]Extractable zinc
[h]Extractable potassium
[i]Extractable iron
[j]Un-inoculated control experiment

Except for pH, EC, and salt contents, where some treatments resulted in reduced contents and others, increased contents in rhizospheric soils of the treated potato plants, the quantities of the rest of the soil properties increased as a result of the rhizobacterial treatments. Although no significant differences ($p = 0.077$) were noted among the potato rhizobacterial treatments and the un-inoculated control with regards to Fe contents of the rhizospheric soils, all rhizobacterial treatments resulted in increased Fe contents of between 99.0% and 99.8%. The effects of different rhizobacterial treatments on the nutrient concentration in potato tubers in the screen house experiment are provided in Table 5.

Table 5 Effects of rhizobacterial treatments on physicochemical properties of potato nutrient contents

Treatment	Nitrogen (%)	Protein (%)	Phosphorus (mg kg^{-1})	Potassium (mg kg^{-1})	Iron (mg kg^{-1})	Zinc (mg kg^{-1})
S1	1.05 (87.6) a	6.55 (87.1) a	2363.13 (93.1) a	1051.29 (53.7) bc	0.59 (62.7) ab	1946.68 (91.9) cd
S2	0.91 (85.7) a	5.69 (85.4) a	2052.73 (92.1) a	1341.95 (63.8) b	1.07 (79.4) ab	4101.05 (96.2) abc
S3	0.90 (85.4) ab	5.62 (85.2) ab	2028.50 (92.0) ab	1489.33 (67.4) b	0.95 (76.8) ab	1553.80 (90.0) cd
S4	1.00 (86.9) a	6.25 (86.7) a	2256.39 (92.8) a	2741.86 (82.3) a	1.50 (85.3) ab	6429.97 (97.6) a
S5	0.45 (71.11) bc	2.81 (70.5) bc	1012.16 (83.9) bc	1430.58 (66.0) b	6.63 (96.7) a	2580.98 (94.0) bcd
S6	0.99 (86.9) a	6.21 (86.6) a	2240.02 (92.7) a	1266.93 (61.6) bc	0.52 (57.7) ab	4130.45 (96.2) abc
S7	0.34 (61.8) c	2.14 (61.2) c	770.51 (78.9) c	2327.07 (79.1) a	1.97 (88.8) ab	5629.40 (97.2) ab
S8	0.85 (84.7) ab	5.27 (84.3) ab	1901.46 (91.4) ab	982.14 (50.5) bc	0.57 (61.4) ab	872.16 (82.1) cd
S9	0.20 (35.0) c	1.22 (32.0) c	441.12 (61.1) c	740.97 (34.4) bc	3.53 (93.8) ab	1166.55 (86.6) cd
S10	0.83 (84.3) ab	5.16 (83.9) ab	1860.86 (91.3) ab	947.85 (48.7) bc	0.68 (67.7) ab	1653.20 (90.6) cd
Control[a]	0.133 c	0.833 c	162.677 c	486.328 c	0.216 b	155.947 d
p value	< 0.0001	< 0.0001	< 0.0001	< 0.0001	0.050	< 0.0001
Significant	Yes	Yes	Yes	Yes	Yes	Yes

S1: *Serratia marcescens* NGAS9, S2: *Serratia liquefaciens* KIBS5, S3: *Klebsiella oxytoca* MWAKS1, S4: *Enterobacter tabaci* MATS3, S5: *Enterobacter tabaci* LUTS2, S6: *Enterobacter asburiae* MWAKS5, S7: *Citrobacter freundii* MWANS4, S8: *Enterobacter tabaci* MPUS2, S9: *Klebsiella grimontii* LUTS1, S10: *Enterobacter ludwigii* BUMS
Values are means of three replicates with % increase relative to the control in parenthesis. Means with similar letters within the same columns are not significantly different (ANOVA + Tukey's HSD; $p < 0.05$)
[a]Un-inoculated control experiment

All the studied nutrients in the potato tubers were significantly different ($p < 0.05$) across the different rhizobacterial treatments and the un-inoculated control. The greatest N, protein, and P increments (>85%) were observed for the treatments with *S. marcescens* NGAS9, *E. tabaci* MATS3, and *E. asburiae* MWAKS5. For K, *E. tabaci* MWATS 3 yielded the highest average of 2741.86 mg kg^{-1} corresponding to an 82.3% increment over the un-inoculated control for which the average K content was 486.33 mg kg^{-1}. Similar results were observed for Fe quantity in potato tubers where the same treatment resulted in tubers with an average Fe content of 6.63 mg kg^{-1} corresponding to a 96.7% increment over the control which recorded an average Fe content of 0.126 mg kg^{-1}. Interestingly, the rhizobacterial treatments resulted in tubers with improved Zn content to a great extent over the un-inoculated controls by between 90% and 97%. The highest Zn content of 6429.97 mg kg^{-1} was recorded for the treatment with *E. tabaci* MATS3, an increment of 97.6% over the control whose tubers had an average Zn content of 155.95 mg kg^{-1}.

Discussions

Cultural, Microscopic, Biochemical, and Carbohydrate Utilization Properties of the Isolates

The rhizobacterial isolated exhibited a broad range of morphological features in terms of their colony forms, indicating their relative diversity. All isolates were Gram-negative, agreeing with other reports that that plant rhizospheres are predominantly colonized by Gram-negative bacterial communities. For instance, in a recent study by Mujahid et al. (2015), up to 90% of all studied rhizobacterial isolates from various crop fields, respectively, were also Gram-negative.

Only 3 out of the 10 potato rhizobacterial isolates in this study did not exhibit any form of motility in the MTM. Rhizobacterial motility is an important property that enables bacteria to reach the plant root exudates and flagella-driven chemotaxis is very critical for successful root colonization (Turnbull et al. 2001). Half of the isolates were MR-positive, indicating their ability to produce organic acids which are important in the solubilization of inorganic P (Adeleke et al. 2017). Similarly, the rhizobacterial isolates were all positive for catalase production and some isolates exhibited very strong catalase activities. Catalases are enzymes that act as defense mechanisms for bacteria to detoxify, neutralize, repair, or escape oxidative damages and bactericidal effects of reactive oxygen species like H_2O_2 (Mumtaz et al. 2017). Except for *K. oxytoca* MWAKS1, all the isolates also exhibited citrate utilization which is thought to play a significant role in competitive root colonization and maintenance of bacteria in the rhizosphere (Turnbull et al. 2001).

Half of the potato rhizobacterial isolates exhibited H_2S production. The reduction of sulfide and other sulfate compounds into H_2S is thought to diminish sulfur availability in the soil for plants and is thus not a desirable trait for soil fertility (Choudhary et al. 2018). The isolates all exhibited indole-production in tryptophan-amended cultures, showing their corresponding abilities to produce tryptophanases (Das et al. 2019). Similarly, the rhizobacterial isolates were all positive for the O-F test, indicating their saccharolytic nature which is an important trait for rhizosphere colonization.

The rhizobacterial isolates exhibited varying capacities to metabolize different CHO. A number of them were capable of metabolizing all the sugars but some could not metabolize lactose, sorbitol, and dulcitol. The rhizosphere is generally a nutrient-rich microenvironment due to the presence of rhizodeposits and root exudates with different chemical compositions (Kumar et al. 2018). This can explain the diverse ability of the isolates to utilize different substrates for growth as may be provided for in their natural environments. Substrate preference may confer certain selective advantages in the rhizosphere and multisubstrate utilization may enable rhizobacteria to diversify their nutrient sources for efficient rhizosphere colonization (Zahlnina et al. 2018).

In vitro Plant Growth-Promoting Activities of the Potato Rhizobacterial Isolates

The potato rhizobacterial isolates exhibited varying P, Zn, and K solubilization capacities. The average quantities of solubilized P ranged from 60.96 to 163.47 mg mL^{-1} which is higher compared to previously reported averages for potato rhizobacterial isolates, for example, in studies by Naqqash et al. (2016) where the averages ranged from 30.71 to 141.23 mg L^{-1}. The best P solubilizers were *E. tabaci* LUTS 2, *E. tabaci* MPUS2, and *S. liquefaciens* KIBS5 with average quantities of 115.88, 112.59, and 117.43 mg L^{-1} of solubilized P, respectively. The solubilization of P is proposed to occur through acidification by organic acids and results in the production of di- and mono-basic phosphates which are the only plant-available P forms (Awais et al. 2019). The production of organic acids by the rhizobacterial strains in the present study was evidenced in the biochemical assays and can explain their P solubilization abilities and illustrate how valuable they can be in improving potato P nutrition.

Serratia marcescens NGAS9, with average ZSI of 2.94 and *E. ludwigii* BUMS1, with an average of 130.26 mg L^{-1} of solubilized Zn exhibited the best Zn solubilization abilities in the qualitative and quantitative assays, respectively. Although Zn is a micronutrient, its adequate supply is required for proper potato yields (Vreugdenhil 2007). Since only a small portion of Zn occurs in plant-available forms in most soils, Zn solubilizing bacteria (ZSB) such as the ones identified in the present study have the potential of improving the Zn utilization in potato grown soils (Aloo et al. 2019).

Except for a few isolates, the K solubilization abilities of the potato rhizobacterial isolates followed similar trends to P and Zn solubilization abilities. Two isolates *C. freundii* MWANS4 and *E. ludwigii* BUMS1 particularly showed good K solubilization abilities in the quantitative assays by yielding averages of 112.98 and 125.26 mg L^{-1} of solubilized K, respectively. Evidence suggests that about 98%

of K occurs in soils in fixed forms and only about 2% is available in plant-accessible forms (Meena et al. 2018). As such, efficient KSB such as the ones identified in the present study can significantly enhance potato K nutrition.

All the potato rhizobacterial isolates produced IAA in tryptophan-amended culture media similar to reports by Naqqash et al. (2016). Indole-3-acetic acid is a rhizobacterial PGP hormone that is important for the proliferation of lateral roots and root hairs and enhancement of plant mineral nutrients uptake (Kumar et al. 2018). The average IAA quantity produced by the isolates in the present study was 5.57 ± 4.51 µg mL^{-1}. In a recent study by Jadoon et al. (2019) in Pakistan, lower IAA average quantities of only 2.09 µg mL^{-1} were reported but geographical differences could explain this variation. The isolates generally produced lesser quantities of GA with an average of only 0.423 ± 0.420 µg mL^{-1}. The best GA producer was *E. tabaci* MATS3 with an average of 1.27 µg mL^{-1}. Unlike IAA, reports on rhizobacterial GA production are scanty (Amar et al. 2013), yet GA production is one of the rhizobacterial PGP mechanisms (Aloo et al. 2019).

The average N_2-fixation zones and quantities of NH_3 were 1.153 ± 0.440 cm and 27.97 ± 21.09 mg L^{-1}, respectively. *Serratia marcescens* NGAS9, with an average N_2 fixation zone of 1.70 cm and *E. ludwigii* BUMS1, with an average NH_3 of 79.84 mg L^{-1} yielded the best results in this assay. The diazotrophic abilities of the potato rhizobacteria established in the present investigation indicate the critical role they could be playing in the potato rhizosphere. Although diazotrophy is a common trait in legume symbioses, nitrogenase genes are present in diverse bacterial taxa (Gyaneshwar et al. 2011). Such traits can be optimized and exploited to promote N nutrition in nonlegumes such as the potato using the diazotrophic strains identified in the present study.

The potato rhizobacterial isolates were all capable of producing siderophores which are important metabolites with a high affinity for binding Fe and promoting its availability to plants (Mhlongo et al. 2018). Interestingly, the present isolates showed higher siderophore production abilities than has been reported in other studies for potato rhizobacteria. For instance, the average SU obtained for the isolates in the present investigation was $26.14 \pm 18.25\%$ while in studies by Pathak et al. (2019), potato rhizobacteria produced lower SU means ($< 11.97\%$). Very few potato rhizobacteria have been associated with the siderophore production trait (Aloo et al. 2019), and these siderophore-producing rhizobacterial isolates are important candidates for potato biofertilization.

Effects of the Rhizobacterial Treatments on Growth and Yield of Potted Potatoes

The present study also evaluated the effects of indigenous rhizobacterial treatments on various growth parameters of potted potato under screen house conditions. The results showed that most of the rhizobacterial treatments reduced the DTE and DTF of the potato plants by up to 7.35% and 32.68%, respectively, relative to the un-inoculated controls. Increased germination rates and seedling vigor in plants following inoculation with beneficial rhizobacterial strains are advanced to occur as a result of phytohormone production that enhances growth by stimulating root elongation and development (Ahemad and Kibret 2014).

Except for the number and weight of tubers in the present study, the rest of the potato growth parameters were not significantly different across the treatments and the control treatment. Nevertheless, the rhizobacterial treatments still resulted in increased growth attributes of the plant. For instance, the potato shoot weights were increased by 22–88% upon rhizobacterial inoculation. Such results can also be attributed to the stimulation of root development and nutrient uptake by rhizobacterial PGP hormones (Kumar et al. 2018), whose production was also established for the present rhizobacterial inocula. Contrary to the expectation, *E. ludwigii* BUMS resulted in average potato shoot length and weight that were less than those of the un-inoculated control by 3.94% and 5.94%, respectively. The failure of rhizobacterial inocula to produce the desired results during *in planta* investigations is probably due to the inabilities to establish themselves in the rhizosphere (Istifadah et al. 2018).

The potato rhizospheric soils were also greatly influenced by the rhizobacterial treatments. Most treatments resulted in reduced pH levels relative to the control and increased N, P, K, Zn, and Fe contents in the potato rhizospheres, signifying rhizosphere acidification which is commonly associated with the solubilization of nutrients in the soil. The increased availability of N and P in the rhizospheric soils may be attributed to N_2 fixation and P solubilization by the rhizobacterial inocula as advanced by Sood et al. (2018). The Fe contents in the potato rhizospheric soils increased by up to 99.7% relative to the un-inoculated control following rhizobacterial inoculation, signifying the excellent Fe-mobilization abilities by the rhizobacterial inocula. The soil OC and OM contents also increased significantly for most of the treatments relative to the un-inoculated control.

The present study established that most of the rhizobacterial treatments resulted in tubers with increased nutrient contents, demonstrating improved nutrient uptake and accumulation by the treated plants. This can mostly be attributed to the multitrait inoculants used to treat the potato plants. For instance, the increased uptake and accumulation of N and P may have been due to increased fractions of the minerals in the rhizospheric soils mediated by the rhizobacterial treatments through N_2 fixation and P solubilization, respectively, as similarly observed in wheat by Sood et al. (2018). The inoculation of seed potato tubers with *S. marcescens* NGAS9, *S. liquefaciens* KIBS5, and *E. asburiae* MWAKS5 resulted in tubers with significantly higher N and protein contents, a clear indication of their efficient diazotrophic roles. Interestingly, the treatment of potato seed tubers with *C. freundii* MWANS4 and *K. grimontii* LUTS1, despite exhibiting N_2 fixation abilities in the *in vitro* studies, did not lead to significant increments on the average concentration of N and protein in the potato tubers relative to the un-inoculated control, probably due to the inability to establish adequately themselves in the potato rhizosphere.

Conclusions

The study establishes the importance of indigenous rhizobacterial communities in the biofertilization of potato which can be exploited for its sustainable cultivation. The selected potato rhizobacterial isolates demonstrated efficient N_2-fixing, P-solubilizing, and IAA, siderophores producing abilities. All these characteristics are important PGP traits and have been found effective in positively improving the growth of potted potato plants under screen house conditions. In sustainable crop production, the focus should not only be on increasing crop productivity but also the nutritional value of the food produced for food security. Apart from improving the potato growth parameters relative to the control, the rhizobacterial treatments also enhanced nutrient availability in the rhizospheric soils and improved the potato tuber nutrient contents. The studied isolates are, therefore, potential candidates in future field applications and sustainable cropping of potato in Tanzania.

References

Adeleke R, Mwangburuka C, Oboirien B (2017) Origins, roles and fate of organic acids in soils: a review. S Afr J Bot 108:393–406. https://doi.org/10.1016/j.sajb.2016.09.002

Ahemad M, Kibret M (2014) Mechanisms and applications of plant growth promoting rhizobacteria: current perspective. J King Saud Univ-Sci 26:1–20. https://doi.org/10.1016/j.jksus.2013.05.001

Aloo BN, Mbega ER, Makumba BA (2019) Rhizobacteria-based technology for sustainable cropping of Potato (*Solanum tuberosum* L.). Potato Res:1–21. https://doi.org/10.1007/s11540-019-09432-1

Aloo BN, Mbega ER, Makumba BA, Hertel R, Danel R (2020) Molecular identification and in vitro plant growth-promoting activities of culturable Potato (*Solanum tuberosum* L.) rhizobacteria in Tanzania. Potato Res. https://doi.org/10.1007/s11540-020-09465-x

Amar JD, Kumar M, Kumar R (2013) Plant growth promoting rhizobacteria (PGPR): an alternative of chemical fertilizer for sustainable, environment friendly agriculture. Res J Agric For 1:21–23

AOAC (1995) Official methods of analysis, 16th edn. Association of Official Analytical Chemists, Washington, DC

Awais M, Tariq M, Ali Q, Khan A, Ali A, Nasir IA, Husnain T (2019) Isolation, characterization and association among phosphate solubilizing bacteria from sugarcane rhizosphere. Cytol Genet 53:86–95. https://doi.org/10.3103/S0095452719010031

Batista BD (2012) Promoção De Crescimento Em Milho (*Zea mays* L.) Por rizobactérias associadas à cultura do guaranazeiro (*Paullinia cupana* var. sorbilis). Dissertation, University of Sao Paulo

Bremmer JM, Mulvaney CS (2015) Nitrogen total. In: Page AL (ed) Methods of soil analysis, 2nd edn. American Society of Agronomy, Madison, pp 595–624

Chaurasia S, Gupta AD (2014) Handbook of water, air and soil analysis. International Science Congress Association, Madhya Pradesh

Chi CM, Wang ZC (2010) Characterizing salt-affected soils of Songnen plain using saturated paste and 1:5 soil to-water extraction methods. Arid Land Res Manag 24:1–11. https://doi.org/10.1080/15324980903439362

Choudhary M, Ghasal P, Yadav R, Meena V, Mondal T, Bisht J (2018) Towards plant-beneficiary rhizobacteria and agricultural sustainability. In: Meena VS (ed) Role of rhizospheric microbes in soil: volume 2: nutrient management and crop improvement. Springer Nature, Singapore, pp 1–46

Das S, Nurunnabi TR, Parveen R, Mou AN, Islam ME, Islam KMD, Rahman M (2019) Isolation and characterization of indole acetic acid producing bacteria from rhizosphere soil and their effect on seed germination. Int J Curr Microbiol Appl Sci 8:1237–1245

Edi-Premona M, Moawad MA, Vlek PLG (1996) Effect of phosphate- solubilizing *Pseudomonas putida* on growth of maize and its survival in the rhizosphere. Indones J Crop Sci 11:13–23

FAO (2008) International year of the potato: new light on a hidden treasure. Food and Agriculture Organization of the United Nations, Rome

Fasim F, Ahmed N, Parsons R, Gadd GM (2002) Solubilization of zinc salts by a bacterium isolated from the air environment of a tannery. FEMS Microbiol Lett 213:1–6

George H, Ed H (2011) A summary of N, P, and K research with tomato in Florida. The Institute of Food and Agricultural Sciences, University of Florida, Gainesville

Goswami D, Dhandhukia PC, Patel P, Thakker JN (2014) Screening of PGPR from saline desert of Kutch: growth promotion in Arachis hypogea by *Bacillus licheniformis* A2. Microbiol Res 169: 66–75. https://doi.org/10.1016/j.micres.2013.07.004

Gyaneshwar P, Hirsch AM, Moulin L, Chen WM, Elliot GN, Bontemps C, Santos PE, Gross E, Reis FB Jr, Sprent JI, Young PW, James EK (2011) Legume-nodulating betaproteobacteria: diversity, host range, and future prospects. Mol Plant-Microbe Interact 24:1276–1288. https://doi.org/10.1094/mpmi-06-11-0172

Holbrook AA, Edge WLW, Bailey F (1961) Spectrophotometric method for determination of gibberellic acid in gibberellins. ACS, Washington, DC

Hopkins BG, Hornek DA, MacGuidwin AE (2014) Improving phosphorus use efficiency through potato rhizosphere modification and extension. Am J Potato Res 91:161–174. https://doi.org/10.1007/s12230-014-9370-3

Hugh R, Leifson E (1953) The taxonomic significance of fermentative versus oxidative metabolism of carbohydrates by various gram negative bacteria. J Bacteriol 66:24

Istifadah N, Pratama N, Taqwim S, Sunarto T (2018) Effects of bacterial endophytes from potato roots and tubers on potato cyst nematode (*Globodera rostochiensis*). Biodiversitas 19:47–51. https://doi.org/10.13057/biodiv/d190108

Jackson ML (1973) Soil chemical analysis, 2nd edn. Prentice Hall of India, New Delhi

Jadoon S, Afzal A, Asad SA, Sultan T, Tabassum B, Umer M, Asif M (2019) Plant growth promoting traits of rhizobacteria isolated from Potato (*Solanum tuberosum* L.) and their antifungal activity against Fusarium oxysporum. J Anim Plant Sci 29:1026–1036

Kumar A, Patel JS, Meena VS (2018) Rhizospheric microbes for sustainable agriculture: an overview. In: Meena V (ed) Role of rhizospheric microbes in soil. Springer Nature, Singapore, pp 1–31

Manzira C (2011) Potato production handbook. Potato Seed Association, Harare

Meena SV, Maurya BR, Meena SK, Mishra PK, Bisht JK, Pattanayak A (2018) Potassium solubilization: strategies to mitigate potassium deficiency in agricultural soils. Glob J Biol Agric Health Sci 7:1–3

Mhlongo MI, Piater LA, Madala NE, Labuschagne N, Dubery IA (2018) The chemistry of plant-microbe interactions in the rhizosphere and the potential for metabolomics to reveal signaling related to defense priming and induced systemic resistance. Front Plant Sci 9:112. https://doi.org/10.3389/fpls.2018.00112

Miller RO (1995) Extractable chloride nitrate orthophosphate and sulfate-sulfur in plant tissue: 2% acetic acid extraction. In: Kalra YP (ed) Handbook of reference methods for plant analysis. CRC Press, Boca Raton, pp 115–118

Mujahid YT, Subhan SA, Wahab A, Masnoon J, Ahmed N, Abbas T (2015) Effects of different physical and chemical parameters on phosphate solubilization activity of plant growth promoting bacteria isolated from indigenous soil. J Pharm Nutr Sci 5:64–70

Mumtaz MZ, Ahmad M, Jamil M, Hussain T (2017) Zinc solubilizing *Bacillus spp*. potential candidates for biofortification in maize. Microbiol Res 202:51–60

Naqqash T, Hameed S, Imram A, Hanif MK, Majeed A, Van Elsas JD (2016) Differential response of potato toward inoculation with taxonomically diverse plant growth promoting rhizobacteria. Front Plant Sci 7:144. https://doi.org/10.3389/fpls.2016.00144

Pathak D, Lone R, Khan S, Koul KK (2019) Isolation, screening and molecular characterization of free-living bacteria of potato (*Solanum tuberosum* L.) and their interplay impact on growth and production of potato plant under Mycorrhizal association. Sci Hortic 252:388–397. https://doi.org/10.1016/j.scienta.2019.02.072

Payne SM (1993) Iron acquisition in microbial pathogenesis. Trends Microbiol 1:66–69. https://doi.org/10.1016/0966-842X(93)90036-Q

Pradhan P (2016) KOH mount preparation: principle, procedure and observation. Microbiology and Infectious Diseases. http://microbesinfo.com/2016/05/koh-mount-preparation-principle-proce dure-and-observation/. Accessed 28 Apr 2018

Sambrook J, Russell DW (2001) Molecular cloning: a laboratory manual, 1st edn. Cold Spring Harbor Laboratory, New York

Schwyn B, Neilands JB (1987) Universal chemical assay for the detection and determination of siderophores. Anal Biochem 60:47–56

Simmons JS (1926) A culture medium for differentiating organisms of typhoid-colon aerogenes groups and for isolation of certain fungi. J Infect Dis 39:209

Sindhu SS, Gupta SK, Dadarwal KR (1999) Antagonistic effect of *Pseudomonas spp.* on patho-genic fungi and enhancement of plant growth in green gram (*Vigna radiata*). Biol Fertil Soils 29: 62–68

Somasegaran P, Hoben HJ (1994) Handbook for rhizobia. Methods in legume–rhizobium technol-ogy. Springer, Heidelberg

Sood G, Kaushal R, Chauhan A, Gupta S (2018) Indigenous plant-growth-promoting rhizobacteria and chemical fertilisers: impact on wheat (*Triticum aestivum*) productivity and soil properties in North Western Himalayan region. Crop Pasture Sci 69:460–468

Tedesco MJ, Gianello C, Bissani CA, Bohnen H, Volkweiss SJ (1995) Analysis of soil, plants and other materials, 2nd edn. Federal University of Rio Grande do Sul, Porto Alegre

Tran TS, Simard RR (1993) Mehlich III – extractable elements. In: Carter MR (ed) Soil sampling and methods of analysis, 2nd edn. CRC Press, Boca Raton, pp 43–49

Turnbull GA, Morgan JAW, Whipps JM, Saunders JR (2001) The role of bacterial motility in the survival and spread of *Pseudomonas fluorescens* in soil and in the attachment and colonisation of wheat roots. FEMS Microbiol Ecol 36:21–23. https://doi.org/10.1111/j.1574-6941.2001.tb00822.x

Vincent JM (1970) A manual for the practical study of the root-nodule Bacteria. Blackwell, Oxford, UK

Vreugdenhil D (2007) Potato biology and biotechnology – advances and perspectives. Elsevier, Amsterdam

Wahyudi AT, Astuti RP, Widyawati A, Meryandini AA, Nawagsih AA (2011) Characterization of *Bacillus sp.* strains isolated from rhizosphere of soybean plants for their use as potential plant growth promoting rhizobacteria. J Microbiol Antimicrob 3:34–40

Walkley AJ, Black IA (1934) Estimation of soil organic carbon by the chromic acid titration method. Soil Sci 37:29–38

Zahlnina K, Louie K, Hao Z, Mansoori N, Da Rocha UN, Shi S, Cho H, Karaoz U, Loqué D, Powen BP, Firestone M, Northern TR, Brodie EL (2018) Dynamic root exudate chemistry and microbial substrate preferences drive patterns in rhizosphere microbial community assembly. Nat Microbiol 3:480

9

Gender and Climate Change Adaptation Among Rural Households in Nigeria

Chukwuma Otum Ume, Patience Ifeyinwa Opata and Anthony Nwa Jesus Onyekuru

Contents

C. O. Ume (✉)
Agricultural and Environmental Policy Department, Justus Liebig University Giessen, Giessen, Germany
e-mail: chukwuma.ume@agrar.uni-giessen.de

P. I. Opata
Department of Agricultural Economics, University of Nigeria, Nsukka, Nigeria
e-mail: patience.opata@unn.edu.ng

A. N. J. Onyekuru
Resource and Environmental Policy Research Centre, Department of Agricultural Economics, University of Nigeria, Nsukka, Nigeria
e-mail: anthony.onyekuru@unn.edu.ng

Abstract

Female- and male-headed rural households have unequal opportunities in climate change adaptation. Efforts in climate change adaptation in regions with deeply entrenched sociocultural norms should also account for the varied gender components of climate change. The broad objective of this study is to integrate gender issues into climate change adaptation thereby distilling lessons and evidence for policymakers on how to approach the necessary transformation of gender relations in climate change interventions. The study employed focus group discussions to uncover the structural factors undermining women's adaptive capacity, thereby making them vulnerable to climate change impacts. In addition to this, in-depth interviews were also conducted. For the in-depth interviews, 27 farmers were sampled using a snowballing method, while four focus groups were carried out differently for male and female farmers. Ten extension personnel and ten representations from the ministry of agriculture were also surveyed using in-depth interviews. Results from the study showed that female farmers in the region were more vulnerable to climate change as a result of the deeply rooted cultural systems and unwarranted assumptions about women. Findings also suggested that women with high adaptive capacity were less vulnerable to climate impacts. We conclude that gender-responsive climate change adaptation is important in achieving balanced relations that will ensure climate resilience in more equitable and nonhierarchical ways.

Keywords

Vulnerability · Stereotypes · Gender mainstreaming · Qualitative methods · Socioeconomic status · Nigeria

Introduction

Climate change is a global phenomenon undermining the efforts towards achieving the Sustainable Development Goals [SDGs]. The Intergovernmental Panel on Climate Change [IPCC] described climate change as the change in global climate patterns which can be identified by variabilities in climate properties over time (IPCC 2014). Climate change has become more threatening not only to environmental quality but also to the fight against poverty, disease, and hunger. This is due to its direct and indirect impacts on agricultural production. While efforts are being made in addressing the causes of climate change through mitigation, building adaptive capacity is particularly important as it will help tackle the current and future impacts of the phenomenon (IPCC 2007). The IPCC defines adaptation as the "adjustment in natural or human systems to a new or changing environment" (IPCC 2007). In other words, climate change adaptation deals with the ability of a system to cushion possible impacts from climate change and to cope with the outcomes. The impact of climate change in the developing nations, according to Chukwuemeka et al. (2018), is mostly felt by the smallholder farmers, farmers that are highly dependent on rainwater and other climate-sensitive input and resources.

The vulnerability of these sets of farmers is expected to be even more severe in Nigeria, where majority of women folk are involved in agriculture for household consumption, yet are highly marginalized and excluded in climate decisions which directly affect them (Osuafor and Nnorom 2014). Gender relations in Nigerian agricultural sector have systematically subordinated women, limiting their access to adaptation information and supports (Uchem 2011). Chavez et al. (2011) describes gender as a social construct that portrays the distinction in roles and opportunities associated with the male and female sexes and the social relations between them. Studies on climate change adaptation by farmers in Nigeria have shown that gender relations and women exclusion in climate decisions adversely affect climate change adaptation efforts, as women contribute 60–80% of food production in the country, mostly for family consumption (Apata 2011; Tersoo 2013; Otitoju and Enete 2016). However, the underlying institutional factors behind these gender issues in Nigeria have not yet been clarified in literature. This gap in literature is what this study intends to fill. Paying attention to these relations is at the core of framing adaptation strategy that will allow farm households build resilience to the impact of climate change in agriculture (MacGregor 2010).

In Africa, literature in climate change adaptation have shown how smallholder farmers suffer high level of vulnerability due to their low adaptive capacity (Okon et al. 2010; Ume 2017). According to FAO (2019), "climate change impact first affects food systems and livelihood groups with a higher level of vulnerability." The IPCC defined climate change vulnerability as the extent to systems, such as geophysical, socioeconomic, and biological systems, which are prone to and incapable of coping with adverse climate change impacts (IPCC 2007). Otitoju and Enete (2016) further noted that among the smallholder farmers in Nigeria, the female farmers are expected to have an even higher level of vulnerability compared to their male counterparts due to their lower adaptive capacities. This, according to them, is due to socioeconomic and institutional factors that undermine their adaptation efforts. According to Eakin and Wehbe (2009), gender transformation – a reassessment of the socioeconomic and institutional factors and relations – established over time, which determines the relationships between the men and women, is therefore imperative if meaningful adaptation effort is to be achieved.

Understanding the gender dimension of climate change adaptation and the underlying socioeconomic factors influencing gender-vulnerability relations among smallholder farmers in Nigeria presents a veritable approach in moving away from the usual incremental adaptation approach to transformational adaptation response. One key importance of transforming gender relations among smallholder farmers is that it is gender that determines the control and ownership of adaptation resources. As stated in FAO (2011), "If women had the same access to productive resources as men, they could increase yields on their farms by 20–30 percent." Gender analysis among smallholder farmers in Africa is necessary for unraveling how best to mainstream gender issues into climate change adaptation plans and policies.

The impact of gender relations on farmers' level of adaptation has been extensively reported in the literature. Findings from literature suggest that these gender issues vary from location to location according to levels of socioeconomic developments (Gafura 2017). As stated in Gafura (2017), these socioeconomic and institutional factors are still unsettled issues requiring further study in order to establish a coherent scholarship in the area.

As women make up the majority of the smallholder farm labor force in Africa, the available studies on climate change adaptation among smallholder farmers in Africa such as Komba and Muchapondwa (2015) have highlighted the need for studies that will link information and evidence of the underlying determinants of gender inequality from expert knowledge and farmer's perception, in order to better understand the factors that undermine efforts in coming up with formidable and gender-balanced adaptation policies. More so, the few available studies were based solely on quantitative analysis of household surveys without any qualitative component (Enete and Amusa 2010; Chukwuemeka et al. 2018; Onyeneke et al. 2018). The implication is that since most of the underlying determinants of gender relations cannot be quantified, the results will lack the deep and contextual information needed for the appropriate policy interventions. This perhaps explains why studies on the gender dimensions of climate change adaptation among smallholder farmers in Nigeria have not been able to sufficiently identify and address the gender relations issues impeding climate change adaptation efforts in the country.

Gender and Vulnerability

The etymology and usage of the word gender show that the concept and its meanings have evolved throughout history. The historical development dates to the late Middle English, when the word was first used to mean "sex of a human being" (Nancy 1989). Recently, it has been conceptualized as a social construct that appropriates certain characteristics such as behaviors, feelings, and attitudes to the male or female individuals in a society (Jrank 2018). What this means is, by knowing a person's gender, we place her or him in a separate social class, and somehow by doing that we judge her or his actions in accordance with our expectations of that social class. According to Jrank (2018), this categorization ultimately leads to the formation of a gendered society – a society with sociocultural expectations of actions deemed appropriate and inappropriate for females and males. This conceptualization led to the emerging theoretical framings on gender roles and stereotypes.

According to Caragliu et al. (2015), gender roles are socially and culturally defined prescriptions and beliefs about the behavior and emotions of men and women. Linda (2017) noted that it is common for different societies to have a set of belief, ideology, and orientation that shapes sociocultural stereotypes among individuals within that society. Caragliu et al. (2015) remarked that it is these gender roles and stereotypes that give rise to the issue of gender identity – a personal perception of oneself which influences her or his status and relationship with other persons in the wider society. Linda (2017) argues that gender roles and stereotypes are determined by institutional factors such as rules, norms, and routines, and this consequently affect household interactions so that expectations from members of a household will be based on gender. Therefore, whereas it is traditionally acceptable for the men to have multiple streams of income, in countries like Nigeria, women, who are traditionally perceived as homemakers or housewives, due to stereotype threats, will only make do with farm produce and resources within the homes. According to Linda (2017), this stereotype threat arises as a result of the women's awareness that they are likely to be judged based on an overgeneralized societal belief about her gender roles.

Literature dealing with the influence of gender on vulnerability to climate change impacts follows three mutually exclusive criteria based on IPCC (2007) analysis. These three elements are: exposure (extent of susceptibility), sensitivity (degree of impact), and adaptive capacity (ability to adjust) to climate change impacts. The World Health Organization (2015) noted that female farmers in developing nations are more exposed to climate variabilities mostly because they make up greater percentage of the world's poor and are mostly charged with the responsibility of gathering firewood for heating and cooking, fetching water for domestic chores, and other activities directly linked to the environment. In terms of sensitivity to climate impacts, Mgbenka and Mbah (2016) noted that 80% of women in many African countries engage in agriculture and are more reliant for their means of support on natural resources that are vulnerable to climate change impacts. Expounding on this fact, Haque et al. (2012) recounted that more than 90% of the fatalities at the 1991 cyclone in Bangladesh were women, and this was due to their social status, limited skills set, and limited mobility. Finally, in assessing adaptive capacity of female-headed households, the RIO+ center in partnership with Food, Agriculture, and Natural Resources Policy Analysis Network conducted a survey in five southern African countries. The results of the survey showed a strong correlation between gender and adaptation level, with women more likely to have low levels of adaptation (Boko 2007). Further analysis showed that women and men in developing nations do not have equal access to climate change adaptation tools in agriculture, with men more likely to have higher access (Onwutuebe 2019). Other studies in various regions in Africa recorded similar results (MacGregor 2010; Ume 2018). Onwutuebe (2019) attributed these gender inequalities to the existing sociocultural norms and institutional factors that limit women's rights and control of resources such as land and financial services. However, none of such studies exploring the role of institutional factors in causing as well as transforming these gender differences have been conducted in Nigeria.

In Africa, studies have also shown that women are highly sidelined in climate decision-making processes, and this in turn affects their access to adaptation information and support in the country (Enete and Amusa 2010; Ume 2018). Enete and Amusa (2010) reviewed journal articles on the influence of socioeconomic characteristics on access to adaptation support in Nigeria in order to ascertain its impact on the agricultural sector of the economy and the implications for economic growth.

They noted, in line with RIO+ survey, that gender strongly influences access to support for climate change adaptation in agriculture in the country. Their work demonstrated the fact that although agriculture employs over 80% of the rural populace, Nigeria is the world's third most vulnerable country in terms of the impact of climate change in agriculture. Their main conclusion from the reviewed literature was that the adaptation differential between women and men undermines the capacity of the agricultural sector to absorb climate change multiple stressors and to maintain function in the face of climate change impacts. Similarly, Rammohan (2016) noted that women's lack of adequate access to adaptation support prevents the agricultural sector from evolving into more desirable configurations that build resilience.

Theoretical Framework

Gender Schema Theory

Gender schema theory as a perceptive theory attempts to explain the process involved in maintaining and transmitting gender-linked qualities from one generation to another, thereby creating gendered individuals in a community (Kendra 2020). Introduced by Sandra Bem in 1981, the theory shows that sex-linked information is mainly transmitted from one generation to another through "schemata," which is a systematic way of allowing some gender-associated information to be more imbibed by individuals within a society. This information consequently shapes the perception the society has about a particular gender in relation to the other gender, which eventually leads to gender stereotyping. The theory argues that the degree to which individuals become gender-stereotyped is influenced heavily by institutional factors and cultural transmissions such as norms, media, school, home training, etc. By institutional factors, the theory refers to configurations in the society that steers people's behavior (norms, routines, and rules). This means that changing any negative stereotype will involve a transformation of existing norms, rules, and routines.

Institutional Theory

The institutional theory attempts to explain the process by which societal configurations or social structures (schemas, rules, norms, and routines) become enshrined as authoritative standards guiding social behavior in society (Dobbin and Vican 2015). The theory attempts to unravel how these structures are formed, accepted, and transmitted over time. Dobbin and Vican (2015) argued that these elements of a social structure (norms, rules, and routines) are perceptive and normative elements, that are enforced and endorsed by individuals and organizations which he referred to as "actors" and transmitted by relational systems at different scales and levels. Since they are imposed and upheld by people, they can be altered by affecting changes in the social structures (Dobbin and Vican 2015). The theory suggests that through socialization, better and appropriate institutional elements in the form of cognitive orientation can be passed on to individuals and organizations, and when it is adopted, it becomes a new pattern or normal behavior. Over time, these new behaviors become resilient and sediment such that individuals begin to act autonomously without recognizing that they are being controlled by the institution.

Context of the Study

Study Area

The study area is Enugu State, Nigeria, that is located in the southeast geopolitical zone of Nigeria. The state was purposively selected because (a) the state is regarded as the capital and policy-making seat of the southeast geopolitical zone, (b) majority of the rural dwellers in this state engaged in small-scale farming, and (c) the state is reported to have experienced marginalization of women in climate change adaptation decision-making (Chukwuemeka et al. 2018). In the national census of 2006, the state has a population of about 3,267,837 (National Bureau of Statistics 2020). Based on the similarities in soil characteristics and by extension meteorological properties, Enugu State is divided into three agricultural zones [AZs]. The zones include Enugu zone, Nsukka zone, and Awgu zone (Fig. 1). The state is in a tropical rain forest zone, with a mean daily temperature of 27 °C and monthly rainfall of 18 mm to uncover the structural factors undermining women adaptive capacity, thereby making them vulnerable to climate change impacts.

Data Analysis

Two sets of qualitative data were generated using in-depth interviews and focus groups. Vulnerability analysis technique was employed to analyze the data generated from the interviews, while content analysis was employed to analyze the data generated from the focus groups.

Following Enete and Amusa (2010), we employed the adaptive capacity approach of the vulnerability index analysis to estimate and compare the vulnerability differential between male- and female-headed households in Enugu State, Nigeria. The justification for adopting this approach is based on the assumption that an increasing adaptive capacity (potential adaptation) will lead to a reduction in vulnerability (Enete and Amusa 2010). The measurement of the adaptive capacity of farm households was determined using sets of variables (vulnerability indicators). Results of this analysis are presented in Fig. 2. The variables that formed the adaptive capacity indicators include:

- Income level (farm income)
- Number of extension visit in the last cropping season

Fig. 1 Map of Enugu State showing the three agricultural zones. (Sources: Chukwuemeka et al. (2018))

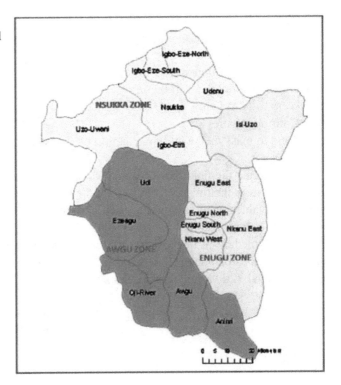

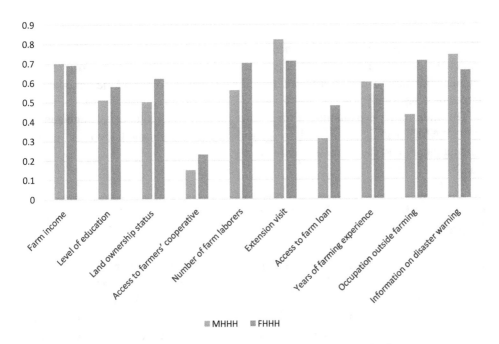

Fig. 2 Gender-vulnerability differential among smallholder farmers in Nigeria

- Educational level
- Land ownership status
- Years of farming experience
- Membership of cooperative societies
- Number of employees in the farm
- Access to farm loan
- Off-farm activities
- Access to mass media and information on disaster warning

To determine the final index, each of the indicators was normalized so that they will be free from their respective units and lie between min = 0 and max = 1. Equal weightage values were assigned to each variable and the average was taken as the value for the vulnerability index. We adopted the United Nations Development Programme method of normalizing life expectancy variables in the Human Development Index calculation (UNDP 2007). The normalization was carried out following Hahn et al. (2009) and Tewari and Bhowmick (2014) using the formula:

$$NSv = \frac{Sv - S\min}{S\max - S\min}$$

where:

S_v = value of the component
NS_v = normalized value for the component
S_{max} and S_{min} = the maximum and the minimum possible values, respectively

By calculating the simple mean of all the variables, the vulnerability index (VI) of all the respondents for the female-headed and male-headed households was determined. To determine the vulnerability index for each gender, we went forward to calculate the mean of the VI for the female-headed households and the male-headed households separately.

The underlining explanation of the gender dimensions in climate change adaptation in Africa can be explained by considering two important theories: gender schema theory and the institutional theory.

Gender-Vulnerability Differential Among Smallholder Farmers in Nigeria

Using years of farming experience as an indicator of adaptive capacity, the average vulnerability index showed that male-headed households (0.60) are more vulnerable compared to female-headed households (0.59). A discussion with the women in the communities showed that women are inducted into the farming occupation quite at an early age compared to their male counterparts. More farming experience suggests that women are better suited to acquire on-farm experience that will enable them autonomously adapt to climate change impacts. For instance, unlike the male farmers, most of the female farmers engaged with could identify several changes in weather patterns they have noticed over the years and how they were able to overcome them. For instance, one of the women stated: "When I was still a youth, working on my mother's farm, we hardly water out maize farms, but now, in other to get a good harvest and fresh grains we do water two or three times in a planting season."

Turning now to off-farm income, female-headed households were found to have a higher vulnerability score. This means that more of the female-headed households rely only on farming for income. We observed that more than half of male-headed households (60%) in the study area reported some off-farming activities such as carpentry, palm wine tapping, and material recycling. The high agriculture dependency index of female-headed households (0.71) suggests that the female-headed households might be very sensitive to climate change impacts, as agriculture, being a very volatile and unpredictable enterprise, is subject to total loss once there is a climate event such as flooding, heatwave, or even disease outbreak. Encouraging the women to engage in off-farm activities, therefore, presents a very good way of reducing their vulnerability.

Both the female- and male-headed households had almost the same VI when information on disasters warning is used as an adaptive indicator (Fig. 2). Information gathered from the Agricultural Development Program (ADP) agents indicated that most of the information on disaster warning is normally conveyed through radio stations. Therefore, only women farmers who have radio sets and are close to radio stations, which are mostly located at Enugu zone, can receive information on disaster warnings and weather forecasts.

In general, when all the vulnerability scores were averaged, findings from the analysis, using the adaptive capacity approach, showed that female-headed households in Enugu State, Nigeria, has lower adaptive capacity than their male-headed households. This finding was in line with previous studies such as Amusa (2010). However, we observed a relatively lower vulnerability gap between female- and male-headed households compared to studies such as Amusa (2010). Studies such as Amusa (2010) used the same variables and adaptive capacity indicators employed in this study and reported (female-headed households = 0.73 and male-headed households = 0.43); however, in this study, we estimated a vulnerability index for the female-headed households = 0.61 and male-headed households = 0.55. However, the variation in result reported can be as a result of variation in study area and time when the studies were conducted. The next section attempts to uncover the institutional factors that influence this gender-vulnerability differential.

Institutional Factors Undermining Climate Change Adaptation Among Female-Headed Households

Most policies and norms that are structured to tackle women's adaptation needs do not address the bigger gender issues. Consistent with the institutional theory, beliefs within the societies have become enshrined as authoritative standards underpinning social justice behavior in the society, irrespective of the negative consequences it carries with it. Some of the cultural practices that need to be transformed are discussed below.

Land Tenure System

This includes the issues of land ownership and transference. From our focus group discussions with the farmers, it was observed that the people hold the belief that if a man dies and he has female and male children, that only the males have the right, by tradition, to possess the family lands. As one of the female farmers stated: *"this 3-story building you are seeing belongs to my brother and the well here for irrigation is also his, these are what he built from his 'Nsukka pepper' farm which is five times the size of mine, and I must pay rent for my own little piece of land."* This is because the culture does not permit women to inherit land, they will resort to renting or saving huge some to purchase their own lands. Scholars in the domain of land reform claim that the main obstacle to climate change adaptation in terms of climate-smart agricultural practices like land rotation, fallowing, etc. is shortage of land and population pressure. However, evidence from our study so far showed that it is not really the shortage of land alone, which affects the adoption of adaptation measures but the structure of land tenure.

Greater population of the farmers in the rural areas are females, but the lack of proper land ownership shifts land title and hence access to the male. Female farmers with both structural and relational access to land have better control and decision on how to put the land to best use. This is predominantly the major problem particularly in most sub-Saharan African countries including Nigeria. The issue of land tenure system, land rights, and climate change adaptation, therefore, presents a very important area that needs to be transformed for better adaptation. Although a strategic gender-responsive and socially inclusive approach can foster a balanced representation of men and women in adaptation initiatives, climate change gender dimensions, and priority to include and empower women and men equally (e.g., by giving them a voice in structural and relations access to land) remain low. Climate change adaptation efforts rest critically on the aspirations and support of those who depend on land-based livelihoods – whose rights and access to land must be protected and promoted for climate change adaptation efforts and initiatives to be sustainable.

Education and Awareness

Similarly, it was observed that it is normal for parents to send their male children to school instead of their female children. The male farmers interviewed were of the opinion that "the females are 'liabilities' but the males are 'assets'". According to one of the male farmers, *"training my female child in school only for her to get married to another man doesn't benefit me."* This shows the level of unjustified assumption about women. Some of these are the assumptions that women farmers

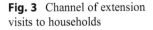
Fig. 3 Channel of extension
visits to households

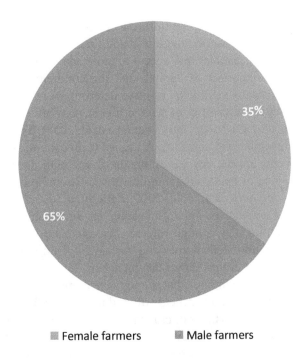

35%

65%

Female farmers   Male farmers

are less informed or intelligent; that some activities are not supposed to be taken up by women; and that women farmers have low self-confidence in taking certain farming decisions. Consistent with the gender-schema theory, these assumptions have been transmitted from one generation to another and have shaped the perception the society has about the female farmers. For instance, it was gathered that extension agents prefer explaining techniques to the male farmers, as they feel they are smarter than their wives, so they can pass the information to their wives. Even the females who oversee the family farms will always project their husband as the household heads, to be better positioned to receive extension services and government aids more easily.

Our study investigated the number of time extension agents visit the female farmers, or gave direct demonstrations to the female farmers, given that extension visits and service delivery helps encourage adaptation. We found that majority of the times (65%), most of the contacts were to the male household heads, though the actual farmers are the wives (Fig. 3).

Several studies (Amusa 2010; Apata 2011; Emeana et al. 2018) have shown the positive influence of direct contact in adoption of climate-smart agricultural practices such as zero tillage (Emeana et al. 2018), organic farming (Apata 2011), and increased soil water conservation practices (Amusa 2010).

Access to Credit and Loan

The result of our analysis revealed that women who have access to credit and loan facilities had a higher possibility of adaptation to climate change. This is in line with the general literature on credit and small-scale farming in the developing nations. According to Otitoju and Enete (2016), access to credits increases farmers coping capacities and access to more adaptation technologies. Similarly, farm size is as important as capital, and the coefficient of farm size was found to be a statistically significant positive determinant of adaptation. This means that females with larger farm sizes have a higher possibility of adapting (Amusa 2010; Enete and Amusa 2010; Chukwuemeka et al. 2018). This also corroborates the economic constraint paradigm which argues that economic constraints and resource endowments are the key determinants of adoption decision such that lack of access to capital or land could significantly constrain both adoption decisions and extent of adoption (Raihan et al. 2009; Zetterlund 2013).

Literature on micro-financing for smallholder farmers shows that microcredit facilities play a huge role in increasing the level of adoption of climate change adaptation technologies and sustainable land-use practices. Emerging theme from our focus groups shows that there exist functional credit schemes, known as "*Akawo Scheme*" in the area that supports the farmers with soft loans. Majority of the farmers explained that the scheme offered them loans from which they were able to support themselves in times of crop failures and even to purchase inputs that they will require in their farms. However, we observed that most of the individuals who did not have access to such networks were females. The female farmers explained that because their husbands are members, they are not allowed to take part as well. Most times, the males are not the farmers, so the loans they get are used for other things but aiding their wives on the farm.

Overcoming the Barriers

We observed two main manners to counter the gender barriers that we illustrated above. First is to improve gender competence and gender training. This means the ability to recognize the social construction and reproduction of gender roles and transform the discriminatory structures and processes. Understanding the gender identities and relations in society will help see the framework conditions provided for women and men, only then can we easily change the perspectives.

Climate change adaptation professionals might be aware of the assessment tools and adaptation strategies; however, they also need to be aware of how they can integrate gender into these strategies and tools in other to acquire gender competence. These can be done in two ways: first is to ensure training on gender and for women. Professionals should be taught on the impact of gender negligence on projects and policy outcomes. Awareness should be raised on how adaptation policies affect men and women differently, making professionals aware of these will have both immediate and long-term impacts. The second is to involve gender experts and specialists that can develop gender-sensitive political measures. They will be able to dictate gender lapses in advance and suggest which areas to pay closer attention to.

The second way is to ensure a bottom-top approach to climate change and gender mainstreaming. One important observation is that the women have indigenous adaptation strategies and are in tune with their environment. They adopt autonomously and can provide strategies that are very beneficial for them. However, adaptation policies are often designed through top-down and bureaucratic models. This has two adverse consequences. The first consequence is that the strategies offered to appear as an imposition of external views and knowledge, and thus appear as unnecessary burden to the female farmers, who are supposed to benefit from them. If the females are engaged in policymaking and are allowed to contribute their knowledge, it would certainly lessen the skepticism and opposition toward adaptation. This will make the policies to focus more on inciting and convincing individuals to autonomously integrate climate impacts prevision in their everyday business. The second consequence of the top-bottom approach is that the male counterparts find it easier to receive the knowledge and adaptation technologies when it eventually drops down to the bottom. If the role of adaptation policy is to reduce vulnerability to climate change through the present modification of the behavior of affected actors so that they integrate and anticipate future impacts of climate change, then it is far from an easy task when the women are engaged actively and at all level in conceiving, planning, and executing climate change adaptation policies.

Conclusion and Recommendation

A gendered perspective on climate change adaptation efforts among smallholder farmers is crucial to sustainable development. This study has added more evidence to the literature on gender and climate change adaptation. Although there appears to be a narrowing of the gender vulnerability gap among smallholder farmers in the area, evidence from this study shows that much work still needs to be done in addressing gender issues in climate change adaptation in Africa, including Nigeria. By taking a gender-sensitive approach, this study highlighted the determinants of adaptation among female-headed households in Nigeria and factors that influence their adaptation to climate change impacts. Furthermore, by incorporating context-specific evidence from the rural farmers, the underlying gender relations and cultural orientations undermining the adaptive capacity of female-headed households were explored. Transforming these gender relations and cultures must form the bedrock for building resilience among farm households in the area. The result of this study shows that cultural systems, policies and practices, and unwarranted assumptions about women are gender relations issues that undermine efforts in building climate change resilience among female-headed households.

Based on the findings from this study, shelving of the identified belief systems that hamper the adaptive capacities of females in the area should be encouraged. It has become obvious that with such societal configurations on the ground, adaptation efforts might be jeopardized. In line with the interactionist theory of gender, since gender relations issues are produced by people through interaction, through a deliberate effort by the people, the unwarranted assumption about women and the enshrined belief systems can also be changed.

References

Amusa TA (2010) Gender and farm household decision on climate change adaptation in southeast, Nigeria. University of Nigeria

Apata TG (2011) Factors influencing the perception and choice of adaptation measures to climate change among farmers in Nigeria. Evidence from farm households in Southwest Nigeria. Available at: https://businessperspectives.org/images/pdf/applications/publishing/templates/article/assets/4402/ee_2011_04_Apata.pdf. Accessed 23 Apr 2020

Boko M (2007) Executive summary Chapter 9: Africa. In: Parry ML et al (eds) Climate change 2007: impacts. London/Cambridge, UK: Cambridge University Press. Print version

Caragliu A et al (2015) Smart cities: transformación digital de las ciudades. In: International encyclopedia of the social & behavioral sciences, 2nd ed, pp 113–117. https://doi.org/10.1016/B978-0-08-097086-8.74017-7

Chavez PR et al (2011) Impact of a new gender-specific definition for binge drinking on prevalence estimates for women. Am J Prevent Med 40(4):468–471. https://doi.org/10.1016/j.amepre.2010.12.008

Chukwuemeka SU, Alaezi K, Ume C (2018) Climate change vulnerability analysis of smallholder farmers in Enugu state Nigeria: gender sensitive approach. J Aridland Agric 4(1):1–6. https://doi.org/10.25081/jaa.2018.v4.3374

Dobbin F, Vican S (2015) Organizations and culture. In: International encyclopedia of the social & behavioral sciences, 2nd ed. Elsevier, pp 390–396. https://doi.org/10.1016/B978-0-08-097086-8.10453-2

Eakin HC, Wehbe MB (2009) Linking local vulnerability to system sustainability in a resilience framework: two cases from Latin America. Clim Change 355–377. https://doi.org/10.1007/s10584-008-9514-x

Emeana E, Trenchard L, Dehnen-Schmutz K, Shaikh S (2018) Evaluating the role of public agricultural extension and advisory services in promoting agro-ecology transition in Southeast Nigeria. Agroecol Sustain Food Syst 43(2):123–144. https://doi.org/10.1080/21683565.2018.1509410

Enete A, Amusa T (2010) Challenges of agricultural adaptation to climate change in Nigeria: a synthesis from the literature. Field Actions Sci Rep 4:1–6. https://journals.openedition.org/factsreports/678

FAO (2011) The state of Food and Agriculture 2010–11. Food and Agriculture Organization of the United Nations, Food and Agriculture Organization. Available at http://www.fao.org/publications/sofa/2010-11/en/. Accessed 23 Apr 2020

FAO (2019) Climate change. Food and Agriculture Organization of the United Nations. Available at: http://www.fao.org/climate-change/en/. Accessed 23 Apr 2020

Gafura AG (2017) Land grabbing, agrarian change and gendered power relations: the case of rural Maasai women of Lepurko village, Northern Tanzania

Hahn MB, Riederer AM, Foster SO (2009) Climate change and human health literature portal the livelihood vulnerability index: a pragmatic approach to assessing risks from climate variability and change – a case study in Mozambique. Glob Environ Change 19(1):74–88. https://doi.org/10.1016/j.gloenvcha.2008.11.002

Haque U et al (2012) Reduced death rates from cyclones in Bangladesh. Bull World Health Organ 90(2):150–156. https://doi.org/10.2471/BLT.11.088302

IPCC (2007) Annex B. Glossary of terms. Intergovernmental Panel on Climate Change. Available at: https://archive.ipcc.ch/pdf/glossary/en.pdf. Accessed 22 Apr 2020

IPCC (2014) Who is who in the Intergovernmental panel on climate change. World Meteorological Organization Building. Switzerland

Jrank (2018) Sex roles – sex-role stereotypes, sex-role socialization – attitudes and behavior, emotional development, Lawrence Kohlberg, and considered – JRank articles. Available at: https://psychology.jrank.org/pages/575/Sex-Roles.html. Accessed 23 Apr 2020

Kendra C (2020) Gender schema theory and roles in culture. Available at https://www.verywellmind.com/what-is-gender-schema-theory-2795205. Accessed 23 Apr 2020

Komba C, Muchapondwa E (2015) Adaptation to climate change by smallholder farmers in Tanzania. Environment for Development Initiative. Available at http://www.jstor.org/stable/resrep15022

Linda L (2017) Gender roles: a sociological perspective – Linda L Lindsey. Google Books. Available at https://books.google.de/books?hl=en&lr=&id=qjjbCgAAQBAJ&oi=fnd& pg=PP1&dpRF1_Sxf93pCuL46fbuqUdbA_4&redir_esc=y#v=onepage&q=genderroles& f=false. Accessed 23 Apr 2020

MacGregor S (2010) "Gender and climate change": from impacts to discourses. J Indian Ocean Reg 6(2):223–238. https://doi.org/10.1080/19480881.2010.536669

Mgbenka RN, Mbah EN (2016) A review of smallholder farming in Nigeria: need for transformation. Int J Agric Ext Rural Dev Stud 3(2):43–54. Available at http://www.eajournals.org/wp-content/uploads/A-Review-of-Smallholder-Farming-In-Nigeria.pdf

Nancy C (1989) Feminism and psychoanalytic theory by Nancy Chodorow: Amazon.com: Books. Available at: https://www.amazon.com/Feminism-Psychoanalytic-Theory-Chodorow. Accessed 23 Apr 2020

National Bureau of Statistics (2020) Pverty statistics. Available at: https://nigerianstat.gov.ng/ elibrary?queries[search]=poverty. Accessed 23 Feb 2020

Okon U, Enete A, Bassey N (2010) Technical efficiency and its determinants in garden egg (Solanum spp) production in Uyo Metropolis, Akwa Ibom State. Field actions science reports. J Field Actions 1(1). Available at: https://journals.openedition.org/factsreports/458

Onwutuebe CJ (2019) Patriarchy and women vulnerability to adverse climate change in Nigeria. SAGE Open. 9(1):215824401982591. https://doi.org/10.1177/2158244019825914

Onyeneke RU et al (2018) Status of climate-smart agriculture in Southeast Nigeria. GeoJournal 83 (2):333–346. https://doi.org/10.1007/s10708-017-9773-z

Osuafor A, Nnorom N (2014) Impact of climate change on food security in Nigeria. Affrev Stech 3:208–219

Otitoju MA, Enete AA (2016) Climate change adaptation: uncovering constraints to the use of adaptation strategies among food crop farmers in South-west, Nigeria using principal component analysis (PCA). In: Tejada Moral M (ed) Cogent food & agriculture, vol 2(1). Informa UK Limited. https://doi.org/10.1080/23311932.2016.1178692

Raihan S, Fatehin S, Haque I (2009) Access to land and other natural resources by the rural poor: the case of Bangladesh. MPRA paper. University Library of Munich, Munich

Rammohan A (2016) Food and nutrition security within the household: gender and access. In: Routledge handbook of food and nutrition security (pp. 368–378). Routledge

Tersoo P (2013) Agribusiness as a veritable tool for rural development in Nigeria. Mediterr J Soc Sci 4(8):17. Available at: https://www.mcser.org/journal/index.php/mjss/article/view/1755

Tewari HR, Bhowmick PK (2014) Livelihood vulnerability index analysis: an approach to study vulnerability in the context of Bihar. J Disaster Risk Stud, 6(1):1–13. Available at: https:// journals.co.za/content/jemba/6/1/EJC163446

Uchem R (2011) Gender roles of men and women in Nigeria and in the United Kingdom. https:// doi.org/10.13140/RG.2.1.2611.3527

Ume C (2017) Critical perspective on climate change adaptation among farmers in developing nations: unpacking divergent approaches. Mod Concepts Dev Agron 1(1). https://doi.org/10. 31031/mcda.2017.01.000504

Ume C (2018) Critical perspective on climate change adaptation among farmers in developing nations: unpacking divergent approaches. Modern Concepts & Developments in Agronomy 1 (1):1–6. https://doi.org/10.31031/mcda.2017.01.000504

UNDP (2007) Human Development Report 2007: Fighting climate change – Human solidarity in a divided world. New York. http://hdr.undp.org/en/content/humandevelopment-report-20078

World Health Organization (2015) Gender, climate change and health. World Health Organization. Available at: https://www.who.int/globalchange.pdf. Accessed 23 Apr 2020

Zetterlund Y (2013) Gender and land grabbing – a post-colonial feminist discussion about the consequences of land grabbing in Rift Valley Kenya. Available at: http://dspace.mah.se/dspace/ handle/2043/15718

10

Impact of Climate Change on Animal Health, Emerging and Re-Emerging Diseases in Africa

Royford Magiri, Kaampwe Muzandu, George Gitau,
Kennedy Choongo and Paul Iji

Contents

R. Magiri (✉) · P. Iji
Fiji National University, College of Agriculture, Fisheries and Forestry, Suva, Fiji
e-mail: royford.magiri@fnu.ac.fj

K. Muzandu
School of Veterinary Medicine, University of Zambia, Lusaka, Zambia

G. Gitau
Faculty of Veterinary Medicine, University of Nairobi, Nairobi, Kenya

K. Choongo
Fiji National University, College of Agriculture, Fisheries and Forestry, Suva, Fiji

School of Veterinary Medicine, University of Zambia, Lusaka, Zambia

Abstract

The threat of climate change and global warming is gaining worldwide recognition. The African continent, because of its size, diversity, and its new status as a "hub" of livestock production, need to gear up to mitigate the possible impacts of climate change on animal health. The aim of this review article is to summarize the current state of knowledge regarding the influence of climate and climate change on the health of food-producing animals. Depending on its intensity and duration, heat stress may directly affect livestock health by causing metabolic disruptions, oxidative stress, and immune suppression, causing increased disease susceptibility, and death. Animal health could also be affected by emergence and re-emergence of vector- and non-vector-borne pathogens that are highly dependent on climatic conditions. The response to these challenges requires community participation in the adaptation of animal production systems to new environments and strengthening the efficiency of veterinary services delivery combined with well-coordinated public health services, since many emerging human diseases are zoonotic.

Keywords

Global warming · Extreme weather · Livestock vulnerability · Adaptation strategies

Introduction

Agriculture serves as the backbone of the economy of most African countries and is the largest domestic income producer. Moreover, it employs about 70–90% of the total working population (Chauvin et al. 2012). Interestingly, this sector supplies up to 50% of household food demand in addition to their income. The livestock subsector provides over half of the value of global agricultural output and one-third in developing countries in Africa. The demand for livestock products in the developing countries is increasing with increasing human population (Steinfeld et al. 2006). However, the livestock subsector is highly susceptible to extreme climate variability. The effect of climate change is anticipated to heighten the susceptibility of livestock production systems to emerging and re-emerging diseases. Climate change is arguably the most important environmental issue currently affecting the livestock sector, but also ecological systems, peoples' livelihoods, and species survival (Zougmoré et al. 2016). Documented effects of climate change are: increasing number of warmer days and nights in a year; changes in rainfall pattern and volume; longer summer seasons; rising sea-levels; and increasing frequency and intensity of floods, droughts, and heat waves (Field et al. 2012; Hughes 2003). Several African governments and regional organizations are aware that increasing demand for livestock products in Africa is not matched by similar growth in local livestock production and are working towards addressing this mismatch in order to reduce dependence on imports.

Unfortunately, there is a lot of emphasis on the impact of climate change on crops in global discussions on climate with little attention paid to livestock production. There is urgent need for a paradigm shift, since it is a fact that even when there is severe crop failure, indigenous livestock have helped vulnerable communities to survive. However, the resilience of indigenous livestock is also threatened by various factors, including extreme climate variability coupled with indiscriminate cross-breeding. With climate change taking the center stage, there is merit in developing indigenous livestock due to their ability to adapt to stressful environments.

The risks associated with unmanaged livestock production in the face of climate change are causing decision-makers in Africa to raise a number of questions such as: What kind of policies would expand livestock production and give societies equal benefits? What is the best way to ensure good health for the people? What choices can ensure that livestock production is socially, biologically, economically, and climatically sustainable? Therefore, the objective of this chapter is to provide some answers to these questions by reviewing the impact of climate change on livestock production, emerging and re-emerging diseases, and adaptation strategies.

Africa's Current Climatic Zones

Climate is a long-term pattern of weather, which is the sum of sunshine, temperature, rainfall, and wind. The amount of heat from the sun plays a significant role in determining climate. The equator cuts across the center of Africa, making it the most tropical continent in the world with different regional climatic conditions (Cooper et al. 2008). The northern to north-eastern parts of the continent and the south-western part of southern Africa have hot and dry climate with unreliable summer rainfall and arid to desert landscape (Gasse 2000). Countries along the equator experience warm and humid climate with heavy summer rainfall for most part of the year and characterized by tropical rain forests. The tropical but not equatorial regions have warm and moist climate with summer rainfall and savanna vegetation. Small areas at the northern and southern tips of Africa have Mediterranean climate, characterized by hot, dry summers and cool, wet winters.

Most of the rain in Africa occurs in the middle of the continent in the north-west to south-east direction (Fig. 1). The influence of cool-drier south-westerly prevailing

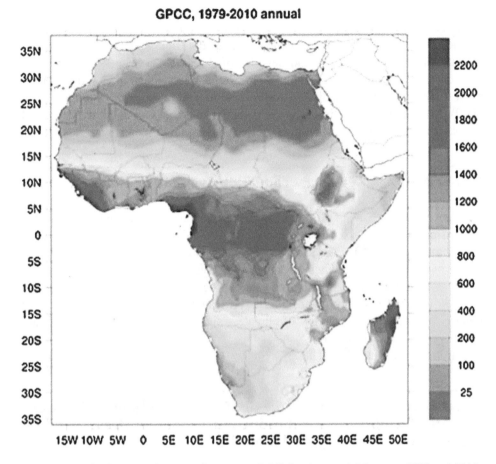

Fig. 1 Map of Africa showing annual average rainfall (mm) averaged between 1979 and 2010. (Siebert 2014)

winds reduces rainfall in the south-western part of Africa while the hot-drier north-easterly trade winds reduce rainfall in the north to north-east Africa.

Expected Impacts of Climate Change on Livestock Production and Health

Africa's contribution to global warming is low while the people living in Africa may be hit hardest by its impact. The overall impact of climate change on animal production and health is far greater than, a mere increase in average annual temperature. The effects of climate change on animal health may be direct or indirect. The direct effects are mainly due to changes in environmental conditions such as air temperature, relative humility, precipitation, drought, and floods (Lacetera 2019). These environmental conditions are responsible for temperature-related animal morbidity and mortality. Indirect effects of climate change on animal health and production are due to microbial density and distribution of vector-borne diseases, food and water shortages, or foodborne diseases (Lacetera 2019). The impact is characterized by: reduced feed quantity and quality; reduced feed intake; water scarcity; increased frequency and severity of diseases and deaths; poor growth rate; decreased quantity and quality of meat and milk; poor reproductive performance; increased costs for disease control and animal production (Thornton et al. 2009). All these culminate into decreased animal health and productivity, decreased animal welfare, and reduced income for affected communities and countries at large.

Vulnerability of Livestock Production Systems to Climate Change

Resilience to the impact of climate change depends on the vulnerability of affected animals and communities. Such vulnerability is the degree to which African livestock production systems are susceptible or incapable of coping with the adverse effects of climate change. In general, developing countries are more vulnerable to the impacts of climate change because they are exposed to many challenges at any time and have limited capacity to adapt. The African livestock sector is particularly vulnerable because of variability of climate, existing disease burdens, culture and economic circumstances, strong dependence on natural resources, weak infrastructure, institutional weaknesses, political and social instability (Thornton et al. 2007).

Direct Effects of Climate Change on Animal Physiology: Metabolic Alterations, Suppressed Reproduction, Oxidative Stress, Immune Suppression, Morbidity, and Mortality

New et al. (2006) analyzed the daily maximum and minimum temperatures and precipitation data from Southern and West African countries for the period 1961–2000. They found a consistent trend of increasing number of hot days and nights and decreasing extremely cold days and nights. Heat stress can occur in animals when a combination of animal and environmental factors increase body heat beyond the normal range (Young 1993). Cattle respond to acute periods of excessive heat in many ways, including decreased feed and increased water intake and increased respiratory rate (Gaughan et al. 1999). On the other hand, Beatty et al. (2006) showed that *Bos taurus* cattle experienced significant physiological changes while similar but less pronounced changes were experienced by *Bos indicus* but without a decrease in feed intake, when exposed to prolonged heat and humidity.

Heat stress due to fluctuating environmental temperatures that exceeds the thermoneutral zone for an animal is the most common stressor that can cause physiological, endocrine and immune responses that may lead to morbidity and reduced animal production (Carroll et al. 2012). Similarly, there was an increase in oxidative stress in dairy cows during the hot season (Tanaka et al. 2007). An increase in oxidative stress has been shown to increase the pathogenesis and severity of many diseases (Halliwell and Gutteridge 1990).

Thus, direct effects of climate change on livestock physiology are observed through the increase in heat stress and have been shown to reduce resistance to diseases and reduce reproductive efficiency and productivity in general. However, heat-tolerant local animals easily adapt compared to heat-susceptible ones.

Pathogen and Vector Ecology

The life cycle of most pathogens involves a free-living stage in the environment while most vectors spend most of their time off the host and only come in contact with animals at the time of feeding.

Pathogen Transmission Ecology and Climate Change

Animal pathogens can live in the body of a host/vector or be free-living in the environment. These pathogens are more susceptible to direct effects of weather change when they are free-living in the environment. The major climate-related factors limiting pathogen survival in the environment are temperature and humidity. Climate change in Africa is causing an increase in the number of days that are hot compared to those that are cold in a year (New et al. 2006). The increasing numbers of hotter days in turn reduce humidity thereby reducing the survival of free-living pathogens in the environment. For instance, the survival of viruses during aerosol transmission or by surface contact is influenced by humidity and ambient temperature (Lowen et al. 2007). Since drought is associated with higher temperatures and low humidity and reduced availability of water, it tends to reduce the survivor and transmission of free-living pathogens into host animals.

Floods increase air humidity and create conditions favorable for survival and transmission of free-living pathogens. For example, the transmission of avian influenza viruses involves the ingestion of water containing the virus (Domanska-Blicharz et al. 2010). Similarly, helminthes and other larger parasites that have a long free-living stage thrive better during weather conditions that increase water availability (van Dijk et al. 2010). Therefore, the effects of climate change on free-living pathogens may take different forms due to fluctuations between drought and flood years. Each pathogen in an area must be assessed on its own merit in order to have a practical adaptation strategy. Furthermore, many pathogens have short generation intervals and high rates of mutation, increasing their ability to evolve over decades and cause emerging diseases (Koelle et al. 2005).

Vector Ecology and Climate Change

In Africa, disease-transmitting vectors are equally affected by increasing number of days with hot weather and low humidity. For vectors whose life cycle involves egg, larval, nymph, and adult stages, the egg and larval stages are the most susceptible to changing weather patterns. It has been shown that the hatchability of *Rhipicephalus decoloratus* eggs reduces when temperatures are low during cold season or flood periods, while larval survival decreases when humidity is low during hot months (Leal et al. 2018). This means that warmer cold seasons and drought increase the population of these ticks and increases the risk of anaplasmosis and babesiosis, while hot seasons with low humidity decrease the population and the risk of these diseases. Similarly, higher temperatures observed during cold seasons (Elbers et al. 2015) and flooding (Ivers and Ryan 2006) may be linked to increased mosquito and Culicoides midge populations, thereby increasing the risk of diseases that they transmit such as lumpy skin Disease, Rift Valley Fever among others.

Vector-Borne Infections as Models of the Effect of Climate Change on Animal Health

There is ample literature on the effects of climate change and the epidemiology of several vector-borne species on animal and human diseases. Vector-borne animal diseases in Africa fall into two main categories: (1) Insect-borne diseases spread by mosquitoes, midges, biting flies, and Tsetse fly which transmit diseases such as Trypanosomiasis and Rift Valley Fever; (2) tick-borne diseases such as East Coast Fever (Bengis et al. 2002).

The mechanisms involved in the emergence of such diseases relating to climate are complex. Vector-borne diseases are usually transmitted by interaction between the hosts and the infected vectors. Infection prevalence is dependent on the interrelationship between hosts, pathogens, and vectors. Any climate-related factor affecting this triangular relationship will affect the vector-transmitted disease epidemiology. In this respect, the survival of the vector, its replication, distribution, density, the vector biting rate, and the pathogen's incubation rate within the vector are of particular importance.

A rise in temperature can allow some vector-borne illnesses to spread in areas with adequate rainfall. Indeed, in the absence of effective veterinary services, the spread of vector-borne diseases is largely determined by a natural boundary where environmental or climatic conditions limit the distribution of the vector and, therefore, the pathogen. Small changes in climatic conditions can have significant repercussions for disease transmission at the fringes of this natural distribution and interfere with endemic stability.

Insect-Borne Diseases

These include (a) arboviral diseases, for example, Rift Valley Fever, Lumpy Skin Disease and African Horse Sickness (b) protozoa diseases, for example, Trypanosomiasis. In the following sections, we will provide evidence, linking climate change to the spread of these diseases.

Mosquito-Borne Diseases

Rift Valley Fever (RVF) is a significant arboviral disease already linked to climate change. It is a peracute or acute zoonotic disease, but affecting mainly domestic ruminants in Africa and transmitted by the mosquito (Mellor and Leake 2000). It is most serious in sheep and goats, resulting in death in newborn animals and abortion in pregnant animals. Epidemics associated with heavy rainfall at irregular intervals of 5–15 years tend to occur in eastern and southern Africa but some outbreaks in Sudan, Egypt, Senegal, and Mauritania are not related to any change in rainfall pattern. Heavy rainfall causes flooding of often dry areas and contributes to the hatching of dormant, drought-resistant, infected Aedes mosquito eggs responsible for the maintenance of the infection in dambos. Emerging mosquitoes infect an amplifying host that becomes a source of infection for many other mosquitoes that spread the disease rapidly.

Rift Valley fever outbursts are accompanied by unexpected heavy rains and flooding, mostly caused by El Niño. For example, the 1997–1998 and 2007 El Niño events were associated with very heavy rainfall in north-eastern Kenya and southern Somalia, resulting in a significant outbreak of RVF (Sang et al. 2010). Rift Valley Fever outbreaks in North and West Africa are not associated with heavy rainfall but with large rivers and dams providing ideal breeding grounds for mosquito vectors. In these areas, increasing the water storage ability for agricultural development and irrigation, in response to the drying climate, is likely to provide new suitable breeding sites for mosquitoes and may make these areas more vulnerable to RVF epidemics. In addition, RVF may become prevalent in areas where the disease has not been previously reported. Wessel born disease (WSL) is another mosquito-borne viral disease of cows, sheep, and goats that largely depends on floodwater breeding Aedes mosquitoes, and has very close epidemiological characteristics to RVF. Lumpy skin disease is a cattle disease of economic importance in Africa, causing severe losses such as damage to hides, mortality, and losses in productivity and reproduction (Krauer et al. 2016).

Midge-Borne Diseases

Wet conditions also promote the breeding of biting flies (e.g., Culicoides (Midges), Stomoxys and Tabanus spp.). African horse sickness (AHS) and bluetongue (BT) viruses are triggered by Culicoides biting midges. Therefore, the dissemination of both diseases is greatly affected by the availability of favorable conditions for breeding of midges. Blue Tongue disease was originally restricted to a belt between 40°N and 35°S worldwide. Its spread to other parts of the world has been attributed to climate change. Some major AHS outbreaks in South Africa have been linked to the combination of drought and heavy rainfall caused by El Niño-Southern Oscillation's (ENSO) warm phase. Many global climate models predict that, in the future, ENSO will occur more often.

Tsetse Fly-Transmitted Diseases

The direct and indirect effects of climate change on tsetse fly distribution and abundance would help to determine the possible spread and prevalence of trypanosomiasis in livestock (Bett et al. 2017). However, a recent study forecasting the expected impact of climate change on the distribution of tsetse flies suggests that the impact of climate change on population results is minimal in comparison with the consequences of population growth and the concomitant shifts in land use and tsetse natural habitat. The biggest shifts in tsetse fly distribution are predicted in the drier areas of western, eastern, and southern Africa (McDermott et al. 2002). Climate change will have less impact on tsetse fly distribution in the tropical regions of Africa. Nevertheless, the potential consequences of increased temperatures and habitat suitability changes on the vector capability of tsetse flies are not clear, and further research is needed.

Tick-Borne Diseases

Diseases which are caused by tick-transmitted pathogens in Africa include: (a) protozoal diseases (Theileriosis and Babesiosis), (b) ricketssial diseases (Anaplasmosis and Heartwater disease), and (c) viral diseases (African Swine fever).

Ticks spend a significant portion of their lives feeding off their host(s) and are therefore prone to atmospheric temperature and humidity. The climate and vegetation influence the environment and determine the distribution and abundance of ticks. Rising temperature, as a result of climate change can shorten their life cycle yet increase their reproductive rate (Estrada-Peña et al. 2012). Rather high temperatures are likely to reduce their longevity, and under drier conditions, mortality may increase. A model developed for the brown-ear tick *Rhipicephalus appendiculatus*, the main vector of East Coast fever (ECF), forecasts that ideal environments for the tick will have vanished in most of the southeastern portion of its distribution by the 2050s. Conversely, in western and central parts of southern Africa, more ideal places for tick survival could appear. A strong correlation between El Niño occurrences and an enhanced ECF seroprevalence has been identified in southern Zambia as a result of the increased tick vector survival (Fandamu et al. 2005). Consequently, temperature strongly affects the nature of the tick population and the diseases they spread by influencing tick distribution and seasonal incidence. Regulation of the big tick-borne diseases in large areas of Africa should focus on preserving an endogenous stable situation.

Therefore, the growth and survival of infectious immunity in bovine with tick-borne diseases is dependent on an optimum interaction between cattle, disease agents, and ticks. Disruption of this optimum relationship due to climate change and subsequent changes in the distribution and availability of certain types of ticks is likely to affect endemic stability and could trigger disease outbreaks.

Non-Vector-Borne Diseases as Models of the Effect of Climate Change on Animal Health

Non-vector-borne diseases are transmitted directly after infectious organisms from the environment enter the animal, usually through inhalation (aerosol) and/or ingestion (water or feed) or through open wounds.

Terrestrial Non-Vector-Borne Diseases

The effect of climate change on the spread of non-vector-borne diseases and their occurrence varies greatly. Changes in environmental conditions may increase or decrease the survival of the infectious agent in the ecosystem or predispose the susceptible animals to infection, as a direct or indirect result of climate change. A changing environment can also lead to increased or decreased interaction between contaminated and vulnerable species, and therefore influence transmission. Changes in temperature and humidity may influence the spatial and temporal distribution of non-vector-borne disease pathogens that spend time outside the host (Van den Bossche and Coetzer 2008). Such infections include: anthrax and blackleg, peste des petits ruminants (PPR), and foot and mouth disease (FMD) found in wind-borne aerosols, dermatophilosis, haemorrhagic septicaemia, coccidiosis, and helminthiasis (Van den Bossche and Coetzer 2008). The increase in rainfall in some areas of Africa creates temporary water bodies or permanent water bodies for irrigation in drier areas in which the intermediate snail host of *F. hepatica* survives (Van den Bossche and Coetzer 2008). While disease spread directly between animals in close contact is less climate-related, changes in the environment resulting in the loss or sporadic availability of water resources or grazing land lead to mass migration of livestock and wildlife in pursuit of water or pasture. Such movement enhances the interaction between livestock from different areas, and between wildlife and livestock, and can contribute to pathogens being transferred. Such population gatherings, animal congregations, and exchange of water and food supplies are considered to significantly contribute to the dissemination of major transboundary African diseases such as FMD, PPR, and contagious bovine pleuropneumonia. Drought, overgrazing, and extreme environmental stress arising from climate change and/or mass migration can become significant trigger factors for soil-borne disease epidemics, such as anthrax which remains dormant and viable in the soil for several decades.

Foot rot is a flood-related bacterial condition that affects ruminant interdigital tissue. Changes in the quality of the interdigital surface, due to exposure to wet conditions during floods, provide a favorable environment for bacterial growth and foot rot (Hiko and Malicha 2016). Changes in the environment as a result of climate change will also affect the migratory paths of a large range of bird species, both within and across continents. These movement path modifications can play a role in the spread and distribution of avian influenza and West Nile virus (Altizer et al. 2011).

Aquatic Non-Vector-Borne Diseases

Epizootic Ulcerative Syndrome (EUS) is a disease characterized by red spots and ulcers on the skin and often high mortality of various fish species (Songe et al. 2011). The lesions of EUS are caused by *Aphanomyces invadans,* and poor water quality parameters predispose fish to EUS (Pathiratne and Jayasinghe 2001). The first outbreak of EUS in Africa was recorded in 2006 in Botswana, Namibia, and Zambia in the Chobe-Zambezi river basin (Andrew et al. 2008). This was associated with

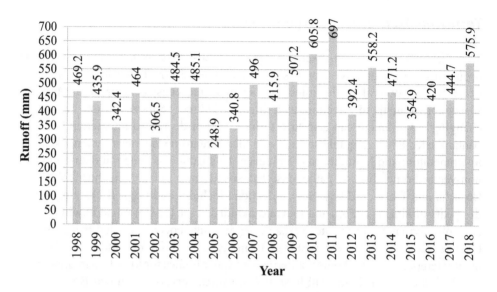

Fig. 2 Total annual water runoff on Zambezi River in Mongu, Zambia from 1998 to 2018 showing the lowest ever water runoff below 250 mm in 2005 due to severe drought. (Source of data, Brakenridge 2020)

river basins having thionic gleysols, soils with enough sulfur content to cause acidification of the soil upon oxidation (FAO/UNESCO 1970), and the severe drought of 2005 (Fig. 1; Brakenridge 2020) that drastically reduced the water table to expose the sulfide/sulfur horizon to air oxidation, resulting in acidic pH and poor water quality (Choongo et al. 2009). The high rainfall of 2006/2007 season flushed the acidic underground water to the surface, creating favorable conditions for the disease. It was observed that the disease was spreading upstream towards the underground source of the water, rather than downstream where all the dead and infected material were being washed to by the water. It is interesting to note that the water runoff on the Zambezi River has up until 2018, never reduced to the 2005 level of less than 250 mm and the EUS disease is so far naturally under control.

Thus, it would be important to use river water runoff data and any other meteorological data that can be used to predict future outbreaks of EUS in areas with underwater sulfur content that can be acidified when exposed to air due to lowering water table during severe drought. This would enable aquaculture farmers in such areas as the Chobe-Zambezi river basin prepare their fish ponds for sufficient alkalinization with lime in order to mitigate the possible outbreak of EUS (Fig. 2).

Determination and Use of Meteorological, Vector and Health Data for Assessing and Mitigating Impacts of Climate Change on Animal Health

The availability of reliable meteorological and disease data is necessary to design effective disease monitoring, surveillance, and early warning and response systems. The major sources of meteorological data, including temperature and precipitation, are obtained from ground-based and satellite measurements or interpolated gridded datasets (Heaney et al. 2016).

A web search for articles published between 1970 and 2015 on meteorology and infectious diseases in Central Africa by Heaney et al. (2016) concluded that in Central Africa, meteorological data is limited because of sparse ground-based recording stations and the lack of proper validation of available satellite data. Therefore, most studies done on the link between infectious diseases and meteorological variability may not be reliable unless infectious disease data are linked to high-quality spatially matched weather data. Spatial mismatch happens when humidity, temperature, and rainfall estimates are obtained from a place that was far from the area where disease data was collected. They advised the scientific community to be aware of the limitations of meteorological data available in Africa and ensure that there is an improvement in the quality and quantity of both meteorological and infectious disease data collected.

"Environmental change vulnerability," which includes community exposure to climatic change, its sensitivity and ability to adapt (McLeman and Hunter 2010), is considered to be very high in African countries. Thus, the need to have quality meteorological and disease data cannot be overemphasized.

Animal and Farmer Resilience and Adaptation to Effects of Climate Change on Animal Health

Coping and adapting to negative effects of climate change on animal health requires that the farming community adapts to changing circumstances, animal management systems are adapted and animal disease preventive measures are strengthened. It should no longer be business as usual.

Community Resilience and Adaptation

Most of the climate-associated emerging and re-emerging animal diseases are transmitted by agents that do not respect any boundaries and therefore, require collective stakeholder participation if effective control for improved productivity is to be achieved. In most African countries, government infrastructure meant to support farmers in livestock production and health is either very poor and unreliable or nonexistent at all. However, when assistance from government or NGOs exist, most of it focuses on the usual aspects of disease control and animal production and not really mitigating effects of climate or weather changes that increase effects of diseases on their animals. Therefore, empowering communities with knowledge and skills to adapt to climatic and weather patterns should now be a priority. Community participatory disease control and surveillance have been practiced in some areas and when supported by government projects or NGOs may yield variable results, which usually disappear when the project or support ends. This is normally the case when project activities are compromised by poor objectives and subjective practicality

(Ayal and Muluneh 2014). In order to have really lasting benefits, there is a need to implement sustainable community based herd health programs that integrate all activities such as vaccinations, deworming, etc., into the day-to-day husbandry, with increased livestock off-take where, there is government marketing support. It has been shown that when farmer livelihood sustainability through improved income becomes the goal of their community participation, tangible results are usually obtained. The rinderpest eradication program was one such success story because livelihood sustenance was the ultimate rationale for collective community efforts (Rich et al. 2014). Therefore, resilience and adaptation to effects of climate change in Africa require that the affected communities are directly involved in adaptation strategies that are practical and centered on livelihood sustenance.

Animal Management Adaptations

Although severe droughts and floods are all considered to be a result of climate change, drought by far has the greatest impact on animal production because it drastically limits both feed and water availability. Limited feed, both in terms of quality and quantity, reduces the competence of the immune system thereby making the animals more susceptible to diseases. In most parts of Africa both feed and water are abundant during flood years. Therefore, in most cases, climate change livestock production adaptation strategies that are usually considered are those responding to the impact of drought. Floods may affect animal production more through diseases than through nutrition. Adaptation strategies can be short-term or long-term.

Unfortunately, most farmers tend to implement short-term adaptation strategies because they have immediate visible impact although they are not usually sustainable in many cases especially for large flocks or herds. Short-term strategies include increasing spending on veterinary services, destocking, trekking animal's long distances in search of water and pasture, constructing fences to protected farmland, as well as feeding through cut-and-carry and other zero-grazing practices.

Long-term resilience to effects of climate change should make management of animals according to local conditions priority number one (Nyamushamba et al. 2017). This should start by ensuring that there are a large percentage of genes from local animals in the breeding stock. Indiscriminate cross-breeding should be discouraged as it produces offspring that are not well adapted and succumb easily to any emerging diseases. Even in the midst of climate change, local breeds have traits for genetic resistance to disease and other traits for quickly adjusting to local conditions. Other management practices like feeding and housing play an important role in the maintenance of resilient, healthy, and productive animals. In fact, continuous breeding of the best performing animals under the harsh local environmental conditions is the more sustainable long-term adaptation strategy. There are many examples of resilient animal breeds that have been developed with mainly local genetics that are performing very well even in the midst of climate change. These include the Tuli and Boran cattle breeds. However, livestock farmers that have maintained their indigenous cattle breeds such as Sanga in Southern Africa (Norval et al. 1988), East

African zebu (Scarpa et al. 2003), and the N'dama in West Africa (Mattioli et al. 2000) suffer less from the impact of environmental changes compared to those that keep breeds with a large percentage of exotic genetics.

Thus management adaptation to changing climate requires committed allocation of resources and improvements in the way land is used for livestock production, coupled with the promotion of local genetics.

Adaptation by Strengthening Disease Prevention Practices

The impact and prevalence of many diseases such as FMD, theileriosis, anaplasmosis, etc., have been increasing while the budget for controlling the diseases has been reducing in most African countries. Sufficient resources must be allocated while ensuring that the whole community is involved in improving sanitation and biosecurity. Usually disease risk factors vary with animal management systems. However, irrespective of the management system, early warning, detection, and quick response are key to preventing and controlling both emerging and re-emerging animal diseases (Black et al. 2008). Prevention and stopping the spread of diseases across an area and country should be the cornerstone of all preventive mechanisms. In addition, in order to stop the spread of high impact infectious diseases such as FMD and ASF, transboundary control measures by African countries are required. The success of the rinderpest eradication program was a result of early detection and early response globally. The world was declared to be free of rinderpest by both FAO and OIE in 2011 because of commitment to early detection and early response (Mariner et al. 2012). Unfortunately, in Africa, new diseases continue to emerge and the impact and persistence of endemic diseases is still a major challenge. In order to achieve early detection and early response, African countries must have functioning laboratories that give real time results of disease confirmation which are well coordinated with field veterinarians.

Conclusion

Climate change has certainly increased land and sea surface temperatures, leading to an increase in the number of warmer days and nights in a year. These changes have been shown to increase the severity of droughts and floods. The manner in which these climatic changes affect animal health and production has been highlighted in this review. The impact can be directly, by affecting the animal's thermoneutral zone thereby inducing stress and its subsequent effects. The effect can also be indirectly by increasing the survival of pathogens and vectors in the environment resulting in increased severity and frequency of diseases. The use of meteorological data that are spatially matched to disease data can help in achieving early warning, detection, and response to eminent diseases. In Africa, adaptation measures should include livelihood-sustenance based community participation in managing animal production according to local conditions and strengthening disease preventive practices.

References

Altizer S, Bartel R, Han BA (2011) Animal migration and infectious disease risk. Science 331(6015):296–302

Andrew T, Huchxermeyer K, Mbeha B, Nengu S (2008) Epizootic ulcerative syndrome affecting fish in the Zambezi river system in Southern Africa. Vet Rec 163:629–632

Ayal D, Muluneh A (2014) Smallholder farmers' vulnerability to climate variability in the highland and lowland of Ethiopia: implications to adaptation strategies. Doctoral thesis, University of South Africa, Geography Department

Beatty DT, Barnes A, Taylor E, Pethick D, McCarthy M, Maloney SK (2006) Physiological responses of *Bos taurus* and *Bos indicus* cattle to prolonged, continuous heat and humidity. J Anim Sci 84(4):972–985. https://doi.org/10.2527/2006.844972x

Bengis R, Kock R, Fischer J (2002) Infectious animal diseases: the wildlife/livestock interface. Rev Sci Tech-Off Int Épizooties 21(1):53–66

Bett B, Kiunga P, Gachohi J, Sindato C, Mbotha D, Robinson T, . . . Grace D (2017) Effects of climate change on the occurrence and distribution of livestock diseases. Prev Vet Med 137:119–129

Black PF, Murray JG, Nunn MJ (2008) Managing animal disease risk in Australia: the impact of climate change. Rev Sci Tech-Off Int Epizooties 27(2):563–580

Brakenridge GR (2020) Global active archive of large flood events. Dartmouth Flood Observatory, University of Colorado. https://floodobservatory.colorado.edu/SiteDisplays/278.htm. Accessed 8 Feb 2020

Carroll JA, Burdick NC, Chase CC Jr, Coleman SW, Spiers DE (2012) Influence of environmental temperature on the physiological, endocrine, and immune responses in livestock exposed to a provocative immune challenge. Domest Anim Endocrinol 43(2):146–153

Chauvin ND, Mulangu F, Porto G (2012) Food production and consumption trends in sub-Saharan Africa: prospects for the transformation of the agricultural sector. UNDP Regional Bureau for Africa, New York

Choongo K, Hang'ombe B, Samui KL, Syachaba MZ, Phiri H, Maguswi C, Muyangaali K, Bwalya G, Mataa L (2009) Environmental and climatic factors associated with epizootic ulcerative syndrome (EUS) in fish from the Zambezi Floodplains, Zambia. Bull Environ Contam Toxicol 83(4):474–478

Cooper P, Dimes J, Rao K, Shapiro B, Shiferaw B, Twomlow S (2008) Coping better with current climatic variability in the rain-fed farming systems of sub-Saharan Africa: an essential first step in adapting to future climate change? Agric Ecosyst Environ 126(1–2):24–35

Domanska-Blicharz K, Manta Z, Smietanka K, Marche S, van den Berg T (2010) H5N1 high pathogenic avian influenza virus survival in different types of water. Avian Dis 54(1 suppl): 734–737

Elbers AR, Koenraadt CJ, Meiswinkel R (2015) Mosquitoes and Culicoides biting midges: vector range and the influence of climate change. Rev Sci Tech 34(1):123–137

Estrada-Peña A, Ayllón N, De La Fuente J (2012) Impact of climate trends on tick-borne pathogen transmission. Front Physiol 3:64

Fandamu P, Duchateau L, Speybroeck N, Marcotty T, Mbao V, Mtambo J, . . . Berkvens D (2005) Theileria parva seroprevalence in traditionally kept cattle in southern Zambia and El Niño. Int J Parasitol 35(4):391–396

FAO/UNESCO (1970) Key to soil units for the soil map of the world. Soil re-sources. Development and Conservation Service, FAO, Rome

Field CB, Barros V, Stocker TF, Dahe Q (2012) Managing the risks of extreme events and disasters to advance climate change adaptation: special report of the intergovernmental panel on climate change. Cambridge University Press, New York

Gasse F (2000) Hydrological changes in the African tropics since the last glacial maximum. Quat Sci Rev 19(1–5):189–211

Gaughan JB, Mader TL, Holt SM, Josey MJ, Rowan KJ (1999) Heat tolerance of Boran and Tuli crossbred steers. J Anim Sci 77:2398–2405

Halliwell B, Gutteridge JM (1990) Role of free radicals and catalytic metal ions in human disease: an overview. Methods Enzymol 186:1–85

Heaney A, Little E, Ng S, Shaman J (2016) Meteorological variability and infectious disease in Central Africa: a review of meteorological data quality. Ann N Y Acad Sci 1382(1):31–43. https://doi.org/10.1111/nyas.13090

Hiko A, Malicha G (2016) Climate change and animal health risk', climate change and the 2030 corporate agenda for sustainable development. Advances in sustainability and environmental justice, vol 19. Emerald Group Publishing Limited, Bingley, West Yorkshire, England, United Kingdom

Hughes L (2003) Climate change and Australia: trends, projections and impacts. Austral Ecol 28(4):423–443

Ivers CL, Ryan ET (2006) Infectious diseases of severe weather-related and flood-related natural disasters. Curr Opin Infect Dis 19:408–414

Koelle K, Pascual M, Md Yunus (2005) Pathogen adaptation to seasonal forcing and climate change. Proc Roy Soc Lond Biol Sci 272(1566):971–977

Krauer F, Riesen M, Reveiz L, Oladapo OT, Martínez-Vega R, Porgo TV, Höfliger A, Broutet NJ, Low N (2016) Zika virus infection as a cause of congenital brain abnormalities and Guillain-Barré syndrome: systematic review: DATASET

Lacetera N (2019) Impact of climate change on animal health and welfare. Anim Front 9(1):26–31

Leal B, Thomas DB, Dearth RK (2018) Population dynamics of off-host Rhipicephalus (Boophilus) microplus (Acari: Ixodidae) larvae in response to habitat and seasonality in South Texas. Vet Sci 5(2):33

Lowen AC, Mubareka S, Steel J, Palese P (2007) Influenza virus transmission is dependent on relative humidity and temperature. PLoS Pathol 3(10):e151

Mariner JC, House JA, Mebus CA, Sollod AE, Chibeu D, Jones BA, Roeder PL, Admassu B, Van't Klooster GGM (2012) Rinderpest eradication: appropriate technology and social innovations. Science 337(6100):1309–1312. https://doi.org/10.1126/science.1223805

Mattioli RC, Pandey VS, Murray M, Fitzpatrick JL (2000) Immunogenetic influences on tick resistance in African cattle with particular reference to trypanotolerant N'Dama (Bos taurus) and trypanosusceptible Gobra zebu (Bos indicus) cattle. Acta Trop 75(3):263–277

McDermott JJ, Kristjanson PM, Kruska R, Reid RS, Robinson TP, Coleman P, ... Thornton PK (2002) Effects of climate, human population and socio-economic changes on tsetse-transmitted trypanosomiasis to 2050. In: The African trypanosomes. Springer, Boston, Massachusetts, USA. pp 25–38

McLeman R, Hunter L (2010) Migration in the context of vulnerability and adaptation to climate change: insights from analogues. Wiley Interdiscip Rev Clim Chang 1:450–461

Mellor P, Leake C (2000) Climatic and geographic influences on arboviral infections and vectors. Rev Sci Tech-Off Int Epizooties 19(1):41–48

New M, Hewitson B, Stephenson DB, Tsiga A, Kruger A, Manhique A, Gomez B, Coelho CAS, Masisi DN, Kululanga E, Mbambalala E, Adesina F, Saleh H, Kanyanga J, Adosi J, Bulane L, Fortunata L, Mdoka ML, Lajoie R (2006) Evidence of trends in daily climate extremes over Southern and West Africa. J Geophys Res 111:D14102. https://doi.org/10.1029/2005JD006289

Norval RAI, Sutherst RW, Kurki J, Gibson JD, Kerr JD (1988) The effect of the brown ear-tick Rhipicephalus appendiculatus on the growth of Sanga and European Breed cattle. Vet Parasitol 30(2):149–164

Nyamushamba GB, Mapiye C, Tada O, Halimani TE, Muchenje V (2017) Conservation of indigenous cattle genetic resources in Southern Africa's smallholder areas: turning threats into opportunities – a review. Asian-Australas J Anim Sci 30(5):603–621. https://doi.org/10.5713/ajas.16.0024

Pathiratne A, Jayasinghe RPPK (2001) Environmental influence on the occurrence of epizootic ulcerative syndrome (EUS) in freshwater fish in the Bellanwila-Attdiya wetlands, Sri Lanka. J Appl Ichthyol 17:30–40

Rich KM, Roland-Holst D, Otte J (2014) An assessment of the *ex-post* socio-economic impacts of global rinderpest eradication: methodological issues and applications to rinderpest control programs in Chad and India. Food Policy 44:248–261. https://doi.org/10.1016/j.foodpol.2013.09.018

Sang R, Kioko E, Lutomiah J, Warigia M, Ochieng C, O'Guinn M, ... Hoel D (2010) Rift Valley fever virus epidemic in Kenya, 2006/2007: the entomologic investigations. Am J Trop Med Hyg 83(2_Suppl):28–37

Scarpa R, Ruto ESK, Kristjanson P, Radeny M, Drucker AG, Rege JEO (2003) Valuing indigenous cattle breeds in Kenya: an empirical comparison of stated and revealed preference value estimates. Ecol Econ 45(3):409–426. https://doi.org/10.1016/S0921-8009(03)00094-6

Siebert A (2014) Hydroclimate extremes in Africa: variability, observations and modeled projections. Geography Compass 8(6):351–367

Songe MM, Hang'ombe MB, Phiri H, Mwase M, Choongo K, van der Waal B, Kanchanakhan S, Reantaso MB, Subasinghe RP (2011) Field observations of fish species susceptible to epizootic ulcerative syndrome in the Zambezi River basin in Sesheke District, Zambia. Trop Anim Health Prod 44:179–183

Steinfeld H, Wassenaar T, Jutzi S (2006) Livestock production systems in developing countries: status, drivers, trends. Rev Sci Tech 25(2):505–516

Tanaka M, Kamiya Y, Kamiya M, Nakai Y (2007) Effect of high environmental temperatures on ascorbic acid, sulfhydryl residue and oxidized lipid concentrations in plasma of dairy cows. Anim Sci J 78(3):301–306. https://doi.org/10.1111/j.1740-0929.2007.00439.x

Thornton PK, Herrero MT, Freeman H, Okeyo Mwai A, Rege J, Jones PG, McDermott JJ (2007). Vulnerability, climate change and livestock-opportunities and challenges for the poor. International Crops Research Institute for the Semi-Arid Tropics (ICRISAT), Patancheru, Hyderabad, India

Thornton PK, van de Steeg J, Notenbaert A, Herrero M (2009) The impacts of climate change on livestock and livestock systems in developing countries: a review of what we know and what we need to know. Agric Syst 101(3):113–127

Van den Bossche P, Coetzer J (2008) Climate change and animal health in Africa. Rev Sci Tech 27(2):551–562

van Dijk J, Sargison ND, Kenyon F, Skuce PJ (2010) Climate change and infectious disease: helminthological challenges to farmed ruminants in temperate regions. Int J Anim Biosci 4(3):377–392. https://doi.org/10.1017/S1751731109990991

Young BA (1993) Implications of excessive heat load to the welfare of cattle in feedlots. In: Recent advances in animal nutrition, Armidale, Australia. Animal Science, Armidale, pp 45–50

Zougmoré R, Partey S, Ouédraogo M, Omitoyin B, Thomas T, Ayantunde A, ... Jalloh A (2016) Toward climate-smart agriculture in West Africa: a review of climate change impacts, adaptation strategies and policy developments for the livestock, fishery and crop production sectors. Agric Food Sec 5(1):26

11

Rainfall Variability and Quantity of Water Supply in Bamenda I, Northwest Region of Cameroon

Zoyem Tedonfack Sedrique and Julius Tata Nfor

Contents

Abstract

Bamenda I municipality found in the humid tropic is endowed with a dense hydrological network which makes it a water catchment for the entire region. Paradoxically, the region still suffers problems of water shortage. This is due to the spatial and temporal variability in rainfall that greatly affects water supply

Z. T. Sedrique (✉) · J. T. Nfor
Department of Geography, Planning and Environment, University of Dschang, Dschang, Cameroon
e-mail: saiyddouk@gmail.com; jtnfor2007@yahoo.com

through its impacts on surface and groundwater. For this reason, we came up with the research topic **"Rainfall variability and quantity of water supply in Bamenda 1, Northwest Region of Cameroon."** The objective of this study is to examine the manifestations of rainfall variability, and how it affects quantity of water supply in the humid tropics. Rainfall data use for this study comprised of annual, monthly, and daily rainfall over a period of 55 years. Water supply data was made of monthly and annual supply. With these data, a Pearson's correlation was computed, and it gave a value of 0.701, with a rainfall proportion of 49.14% and 50.86% for other factors. The seasonality and the Standardized Precipitation Index were equally analyzed. At the end of the study, results showed that rainfall events in Bamenda I fluctuates with time and in space. It equally presented a reduction in the number of rainy days from 204 days in 1663 to 155 in 2018. This led to a reduction in length of rainy season and in rainfall amounts. In addition, the area has witnessed sedimentation of riverbeds and water reservoirs due to erosion and deposition during high rainfall peaks. Equally, floods observed during high rainfall episodes have become a potential threat to water infrastructures imposing exceptional water shortages during the rainy seasons. Due to these, actors in the water supply sector are putting in measures to remedy the situation.

Keywords

Rainfall variability · Water supply · Stream flow · Water catchment · Vulnerability · Adaptation measures · Bamenda I

Introduction

The Intergovernmental Panel on Climate Change (IPCC 2008) relates that there is a consensus that increases atmospheric greenhouse gases and will result in climate change which will cause: rise in sea levels, increased frequency of extreme climatic events including intense storms, heavy rainfall events, and droughts. This will increase the frequency of climate-related hazards on water resources. There is less consensus on the magnitude of change in climatic variables, but several studies have shown that climate change will affect the availability and demand for water resources. Tsalefac (2007) and Wilby et al. (2006) defended the idea that if global warming averagely increases, isotherms will be displaced, leading to a modification of ecosystems and hydrological cycle, and therefore consequent impacts on both surface and groundwater. Climate variability in general and rainfall variability in particular are ills that are affecting mankind today with highland areas being the most vulnerable to the changes in climatic patterns and its related impacts on water resources. As such, this work explores the state of rainfall variability in Bamenda I and its impacts on the quantity of water supply. Our aim is to show how changing rainfall pattern in space and in time as well as extreme weather events such as torrential rains, floods, and droughts have led to disastrous impacts on the water resource and on the water system.

This work analyzes rainfall data from 1963 to 2018 as well as water supply data from 2012 to 2018 with efforts to understand how quantity of water supply has varied with the past and current changes in rainfall patterns. It focuses on the impacts of extreme rainfall events on water supply in the Bamenda I municipality. According to the national plan of action for integrated water resource management (PANGIRE 2013), Cameroon has a dense river network and both surface and groundwater resources are available but not evenly distributed, and both physical and human factors influences the state and availability of its fresh water resources. These factors include climate change, sedimentation of riverbeds, floods, deforestation, physical and chemical contamination, as well as government policies on both waste management and fresh water management. The Bamenda I municipality, found in the humid tropics ought to have had abundant water due to its geographical, geomorphological, geological, and hydrological features. Paradoxically, this region suffers problems of water supply. This is due to the past and current episodes of seasonal droughts and floods caused by heavy and varying rainfall patterns that continue to affects discharge in river basins. The seasonal fluctuations in the quantity of water supply in this locality are equally attributed to the variations in daily, monthly, and annual rainfall variability. These variations have greatly affected the supply of water to the catchment (surface stores), as well as to reservoirs and consequently to different supply points despite the water potentials in the municipality. This work therefore considers rainfall variability as a major constraint to quantity of water supply in the Bamenda I municipality and aims at bringing the impacts of rainfall variability on water supply.

Literature Review

Climate Variability and Water Supply

Much has been said on rainfall (climate) variability and water supply, both in the long term and short term. The impacts of climate on water supply can therefore be seen below.

A group of researchers, Beniston (2003), Sadjin et al. (2005), and Martin Dahinden (2010), who studied mountain and climate change, stand for the idea that mountains are among the most sensitive regions to climate change and that some of the most visible indicators of climate change comes from mountain areas. Their ideas were that mountains are water towers of the world as they provide fresh water to more than half of the world's population, but these areas are also among the most sensitive and vulnerable to climate change.

According to Adefolalu (1993) in his study on precipitation, evapotranspiration, and the ecological zones in Nigeria, he explained the fact that during rainy season, it is not expected to have precipitation on daily basis. However, when breaks of equal or more than three pentads (15 days) occur, they are considered serious anomalies. Also, climate change (prolonged dryness) can lead to the drying up of springs, and it is projected to reduce renewable surface water and groundwater resources significantly in most dry subtropical regions. This will intend to increase the frequency of meteorological droughts (less rainfall) which is likely to increase the frequency of hydrological drought (less surface water and ground water).

Jiduana et al. (2011) explained the variations in the evolution of rainfall patterns and laid emphasis on the fact that the quantity of rainfall and its duration is experiencing a simultaneous decrease in Nigeria which is remarkable to about 78.6% while the intensity of the rainy days and rainy season decreased equally to 77.3%. The study equally permitted to analyze that there has been a regression in the stream flow data over the past years in Nigeria, and this is due to a fall in the quantity of rainfall of about 76.8% affecting the level of stream flow, the stream's transportation capacity, and time of annual recharge. Therefore, it is seen that changes in climate parameters has a potential impact on water supply.

Tsalefac et al. (2007), in their contributions to a book titled *Afrique Centrale, le Cameroun et les changements globaux*, prospected that a certain number of potential consequences of climate change can be seen in a more or less precise manner. If global warming averagely increases, isotherms will be displaced, leading to a modification of ecosystems, mutations in major vegetation types with a great reduction in the forest surface (wood). Drought will increase in the tropical latitudes and thereby increase the risks of extreme weather conditions. This will lead to a reduction in surface and groundwater. Also, certain ecosystems that are very fragile will be particularly sensible to climate change, notably the mountain and coastal ecosystems.

Adaptation to Climate Variability

In the phase of hydrological changes and fresh water-related impacts, vulnerability, and risk due to climate change, there is need for adaptation and for increasing resilience. Managing the changing risks due to the impact of climate change is the key to adaptation in the water sector and risk management should be part of decision-making and treatment of uncertainty (IPCC 2014). To exploit the impacts of climate change on fresh water, adaptation is generally required; there is growing agreement that an adaptive approach to water management can successfully address uncertainty due to climate change.

According to a group of scholars Mark et al. (2008), adaptation strategies to rainfall variability include household water treatment and safe storage (HWTS), water storage and conservation techniques, water reclamation and reuse technics, increasing use of water-efficient fixtures and appliances. These strategies are grouped in to six typology of adaption technologies which are diversification of water supply, groundwater recharge, preparation for extreme weather events, resilience to water quality degradation, storm water control and capture, and water conservation.

Another group of researchers, ML Parry et al. (2007), in water supply sanitation (WSS) propose sector-specific models, which included upgrading existing infrastructure to meet future challenges and cope with the risks associated with climate change. Example, installation of pre-sedimentation pond or riverbank filters for pretreatment, shifting from shallow wells to more reliable sources of water supply, such as surface water and confined aquifers. It was equally concerned with improving water supply through master plan and long-term investment plan including an inventory of groundwater resources. They equally implemented complementary measures, which included the introduction of disaster and climate risk assessment, and improving general framework for risk assessment and management, establishing system for managing floods and other climate and water-related disasters. According to them a proper adaptation could equally be done through protecting water intake facilities from flooding, protecting pumping stations and treatment plants or other facilities potentially exposed to flooding and encouraging the use of alternative water sources, such as (treated) wastewater reuse and rainwater harvesting, to minimize dependence on freshwater and secure access to stable sources of water. Recent experience in Moldova confirms that well properly designed and operated basic treatment plants can deliver even under extreme weather conditions.

Location of Study Area and Research Methodology

Location of the Study Area

Bamenda I subdivision is situated Southeast of Mezam division, one of the seven divisions of the Northwest region of Cameroon. It is located within latitude 5°51" to 5°58" north of the equator and longitude 10°8" to 10° 17" east of the Greenwich meridian, (Fig. 4). It falls within the humid tropical climatic zone and covers a total surface area of about 110 km^2 with one main village (Bamendakwe) comprising of about 13 quarters, namely, Abangoh, Achichum, Alahnting, Aningdoh, Banche, Nkar, Ayaba, Ntanche, Ntafebuh, Nta'afi, Moyo, Keneleri, and Abumuchi. This municipality is situated at the entry to Bamenda town, and it is oriented 366 km northwest of the Cameroon capital with an altitudinal range from about 1269 to 2606 m above sea level.

As seen on Fig. 1, the Bamenda I municipality is bounded to the north by Bamenda III subdivision, to the south by Santa, to the east by Tubah and Balikumbat, and to the west by Bamenda II subdivision (Flyer Bamenda I council 2018). It is connected to the national territory by the national road N°6. Its geographical location falls within the Guinean climate type. This climate varies from one area to another, and it equally varies with altitudes (thermal gradient) and seasons. The climate is marked by two distinct seasons which are the dry and rainy season. The rainy season is usually from around mid-March to mid-October and sometimes extends to November. The rainfall ranges between 2000 and 2500 mm per annum and highest amounts are recorded in July and August. Heavy torrential rainfall in this area usually results from strong Southwest Monsoon winds blowing into the country from the Atlantic Ocean. These rainfall amounts lead to a considerable recharge of the water table and consequently an increase in stream volume, thereby leading to adequate supply of water to the entire municipality.

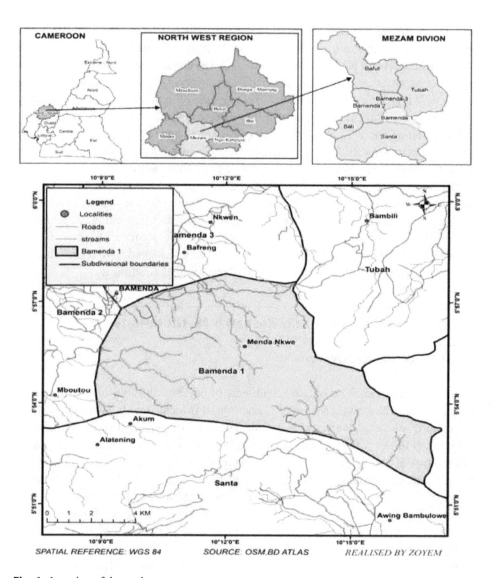

Fig. 1 Location of the study area

On the other hand, dry northeast trade winds from the Sahara Desert blow to the national territory through the northern part of the country, from November to January. The dry season is usually from mid-October and or November to around mid-March. Strong winds and heavy cloud cover characterize the area due to convectional heating. The altitudinal range gives it a highland average temperature of about 19.4 °C with about 18 °C around water catchments, confirming the reduction of temperatures with altitudes. Equally, this area registers average relative humidity of about 76.9%. These moderate temperatures are friendly to water conservation as very little is lost through evaporation. On the other hand, high humidity conditions the availability of moisture in the atmosphere. The area also receives insulations of about 6–8 h/day, reaching its peak in January when the northeast trade winds sweeps across the area. This leads to some cloud formations on the highlands of Mendankwe, and during such occurrences, potential cases of convectional rainfall are recorded. These climatic parameters are favorable for the natural supply of water to the catchments and subsequently to the water system there by qualifying this area as a watershed for the entire locality. It is clearly seen on Fig. 1, as most rivers take their rise from the municipality and flows to different areas.

The area is equally located on the western highland plateau, along the Cameroon dorsal, which runs from Mount Cameroon in the southwest passing through Mount Manengouba and Mount Bamboutos in the west, and stretching to Mount Oku in the northwest. High altitudes with series of mountain chains are common in this area. The locality therefore presents a wide range in its relief forms, with altitudes ranging from 1269 to 2606 m. This range permits an ample downflow of water to the rest of the municipality and beyond. The zone is therefore considered a watershed for the entire region since most rivers and streams draining the region take their rise from its highlands.

Research Methodology

Data Collection

Both qualitative and quantitative data were gotten from respondents in the zone of study, and the data was from primary as well as secondary sources. Primary data was mostly qualitative and comprised of observations made, interviews conducted, and questionnaires administered to both the water management committee and the local population. This information was obtained using questionnaires and interview guides, where 99 questionnaires were successfully administered to seven quarters and returned. Secondary data were both quantitative and qualitative. Quantitative secondary data comprised of the total quantity of water abstracted and supplied in Bamenda I from 2012 to 2018, collected from the Cameroon Water Utility Corporation (CAMWATER), and of rainfall data for Bamenda Station from 1963 to 2018, collected from the chief in charge of the meteorological station at the Northwest Regional Delegation of Transport. These data were used firstly to characterize rainfall events in the Bamenda I municipality, secondly to determine the Standardized Precipitation Index (SPI) and the rainfall seasonality index, and finally the Pearson's correlation coefficient.

Results and Discussions

Manifestations of rainfall variability in Bamenda I was evaluated through analyses of interannual rainfall variability and anomalies, analyses of monthly rainfall variability and anomalies, fluctuation in number of rainy days and in rainfall intensity, and through the application of climatic indices. Also, the Pearson's correlation coefficient was used to evaluate the extent to which manifestations in rainfall affect water supply quantity. These manifestations have consequent impacts on water resource availability or stream discharge in catchments, on riverbanks and on the water systems. This is due to heavy downpours that cause massive erosion in the drainage basin, landslides along river banks, and sedimentation in water reservoirs.

Interannual Rainfall Variability and Anomalies

Rainfall variability in Bamenda I manifest through fluctuations in annual rainfall amounts seen as some years registered more rainfall amounts than others. It equally manifest through annual rainfall anomalies as seen on Figs. 2 and 3, respectively.

As seen on Fig. 2, there have been great variations in annual rainfall amounts over the study period. This is seen as some years registered high rainfall amount (above the means of 2362 mm), while others registered low rainfall amounts (below the mean). High amounts of rainfall where registered in 1963 (2800 mm), in 1969 (2900 mm), in 1981 (2500 mm), and in 2018 (2400 mm), while low rainfall amounts were registered in 1972 (2200 mm), in 1973 (1800 mm), in 2014 (2000), and in 2016 (1600 mm). There have been a general fall in rainfall amount over the past 55 years, indicated by the depressing trend line. This is in line with global trends of falling rainfall with time due to changing climate, IPCC (2008).

On the other hand, Fig. 3 illustrates rainfall anomalies, and it is seen as some years registered positive rainfall anomalies and others negative rainfall anomalies. Example positive anomalies were recorded in years like 1963, 1969, 1976, 1979, and in 2002. They registered anomalies ranging between 300 and 500 mm, while negative anomalies were recorded in 1964, 1973, 2015, 2016, and 2017 with anomalies between 400 and 900 mm. Negative anomalies registered a frequency 60% against 40% for positive anomalies reason for the falling trends.

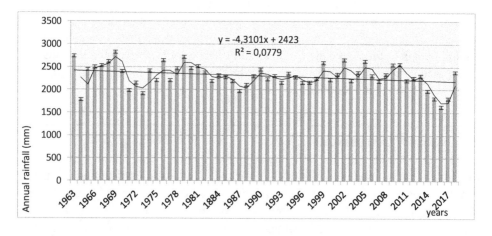

Fig. 2 Interannual rainfall variability in Bamenda up station. (Source: Author (2019))

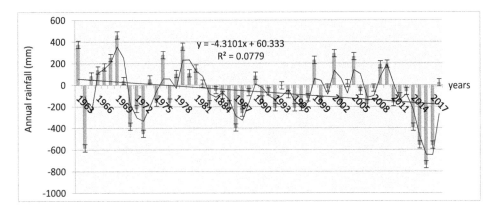

Fig. 3 Interannual rainfall anomalies in Bamenda up station. (Source: Author (2019))

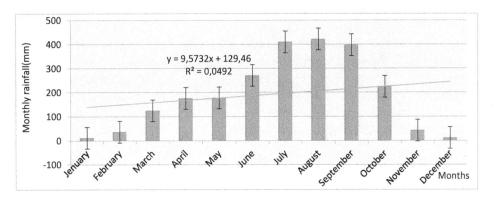

Fig. 4 Monthly rainfall variability. (Source: Author (2019))

Monthly Rainfall Variability and Anomalies

Rainfall variability equally manifests through fluctuations in monthly rainfall amounts and anomalies, but fluctuations follow a seasonal pattern. The mean monthly rainfall was calculated at 191.68 mm and fluctuations went both below and beyond the mean, indicating variations rainfall (Fig. 4).

As seen on Fig. 4, rainfall amounts are lower in months of January with about 10 mm, February with 35 mm, November (42 mm), and December with about 11 mm, as they all register rainfall below the mean of 191.68 mm. The rainfall begins to rise in the months of March (120 mm), which mark the beginning of the rainy season. Its reaches its peak in July (410 mm), and August (420 mm), and starts falling around October (220 mm).

Rainfall variability in the Bamenda I municipality is equally indicated by monthly rainfall anomalies. This is seen as some months record positive rainfall anomalies while others record negative anomalies (Fig. 5).

Figure 5 on the other hand illustrate monthly rainfall anomalies, and it is seen that negative anomalies were recorded in some rainy months (March, April and May).

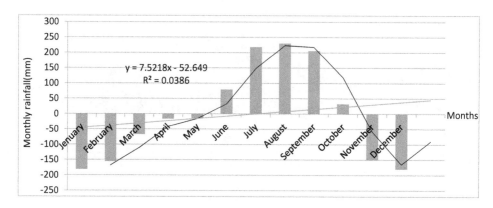

Fig. 5 Monthly rainfall anomalies. (Source: Author (2019))

This is paradoxical because negative anomalies are expected in months of the dry season, but they are instead recorded in rainy months. This may be accounted for by the late onset of rains leading to an increase of the length of the dry season. This will lead to a drop in the water table, thereby reducing water level in catchment and thus exacerbating water shortages. This confirms the results of Hayward et al. (1987) that "an increase in the length of time for recharge of ground water due to prolonged dry season will cause a drop in the water table and a drying up of some streams and springs on which people depend for survival."

Fluctuation in Number of Rainy Days and Rainfall Intensity

Variation in the number of rainy days and changes in rainfall intensity are other indicators of rainfall variability. This variation has a link with fluctuations in the dates of onset and retreat of rains. This link is seen as years with late onset and early retreat will have a higher tendency of registering less rainy days but with higher rainfall intensity while years with early onset and late retreat will have higher chances of recording many rainy days but less rainfall intensities.

Figure 6 shows variability in rainy days in Bamenda up station from 1963 to 2018. This study period has registered a lot fluctuation in number of rainy days as some years registered close to 250 rainy days annually and others about 130 rainy days. Highest number of rainy days were recorded in 1999 (249 days) and in 2007 (225 days). On the other hand, 1964 and 2011 had the least number of rainy days (135 days and 130 days, respectively). The annual difference in number of rainy days is about 119 days, which is much enough to attest that there have been variations in rainfall within the study area. This is because fluctuations in number of rainy days will obviously lead to fluctuations in frequency and intensity of rainfall. This is felt as years with less rainy days and more rainfall amount record high rainfall intensity, while those with more rainy days and less rainfall amount record less rainfall intensity. This is because less rainfall amount is rationed within

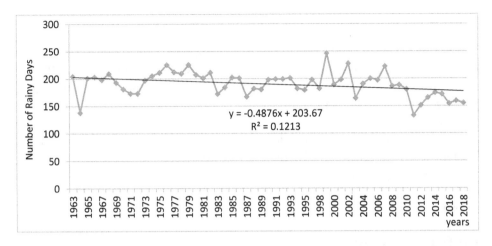

Fig. 6 Number of rainy days. (Source: Author (2019))

Table 1 Classification of SPI values, categories, and corresponding proportions

SPI value	Category	Probability (%)
2.00 or more	Extreme wet	0.3
1.5 to 1.99	Severely wet	2.4
1.00 to 1.49	Moderately wet	11.2
0 to 0.99	**Mildly wet**	**30.1**
0 to −0.99	**Mild drought**	**37.4**
−1.00 to −1.49	Moderate drought	14.2
−1.50 to −1.99	Severe drought	4.4
−2 or less	Extreme drought	0

Source: McKee et al. (1993)

the many rainy days and consequently, low rainfall intensity, whereas with the few rainy days, much rainfall amount is pouring per unit area, and hence high rainfall intensity.

Analyses of Rainfall Variability in Bamenda I Using Climatic Indices

Climatic indices were equally used to characterize rainfall variability in Bamenda I, and both the seasonality and the standardized precipitation index were used (Tables 1 and 2).

The Standardized Precipitation Index (SPI)

The Standardized Precipitation Index (SPI) is a tool used primarily for defining and monitoring drought and floods (rainfall situation). It equally serves as a method of assessing climatic variability. It was developed by McKee et al. in 1993. It was therefore used to access the situation of rainfall variability in Bamenda I (Table 1).

As seen on Table 1, 0.3% of the study period registered SPI values greater than or equal to 2.00 indicating extreme wet conditions. This implies that the probability of occurrence of extreme floods in Bamenda I is 0.3%. Equally, 2.4% of the study period registered values between 1.5% and 1.99% corresponding to severely wet conditions. Also, 11.2% recorded moderately wet and 30.1% recorded mildly wet conditions. To what concerns drought conditions, 37.4% lived mild drought, 14.2% lived moderate drought, and 4.4% lived severe drought conditions, but no situation of extreme drought was witnessed. The SPI therefore indicates that mild wet and mild drought conditions are most common in Bamenda I since they have the highest probability of occurrence.

The Seasonality Index

The seasonality of precipitation refers to the tendency for a place to have more rainfall in certain months than in others. It therefore determines the rainfall regime of a particular place (length of seasons). This makes use of the seasonality index (SI) of Walsh and Lawler (1981). Table 2 presents the seasonality index for Bamenda I.

Table 2 shows the scale (class limit) of the seasonality index and their corresponding seasonal regime. The seasonality index for Bamenda up station is calculated at an overall value of **0.901**, and as seen on the scale, this value falls within the limit **0.80 to 0.99**. This means that the rainfall regime for Bamenda station portrays a markedly seasonal with a long drier season instead of a seasonal regime. Instead of seasonal because humid tropics (Bamenda) are characterized by two distinct seasons, so we should normally have a "seasonal" rainfall regime with a well-defined rainy and dry season but instead of a seasonal regime, it shifted to a "markedly seasonal with long drier season." This is a clear indicator of a variation in rainfall, and it also indicates that rainfall in Bamenda I has been reducing gradually, leading to higher trends of drought occurrence. This result is in line with that of Jiduana et al. (2011), in Northern Nigeria, which emphasizes on the fact that "rainfall amounts and its duration is experiencing a general decrease." This explains why rainy months of March, April, and May portray negative anomalies, with rainfall

Table 2 Rainfall seasonality index and corresponding regimes

Seasonality index class limit	Rainfall regime
≤0.19	Very equable
0.20–0.39	Equable but with a definite wetter season
0.40–0.59	Rather seasonal with a short drier season
0.60–0.79	Seasonal
0.80–0.99	**Markedly seasonal with long drier season**
1.00–1.19	Most rain in 3 months or less
≥1.20	Extreme, almost all rain in 1–2 months

Source: Walsh and Lawler (1981)

below the normal. This fall in rainfall and occurrence of droughts will obviously lead to a fall in the general quantity of water supply in Bamenda I.

Impacts of Rainfall Variability on Water Supply

This section deals with the effects of variation in rainfall on water supply. These variations in rainfall pattern and amount have caused several impacts on the water system. They include intense erosion within river channel leading to occurrence of landslides, sedimentation in water collection chambers leading to its blockage, and erosion along pipeline network leading to breakage. Also, meteorological droughts lead to shrinking of rivers and springs. It equally shows the Pearson's correlation coefficient between monthly rainfall amount and water supply.

Pearson's Correlation Coefficient Between Rainfall Variability and Water Supply

The Pearson's correlation coefficient is a measure of the linear correlation between two variables. It is a tool for determining the strength of a relationship existing between two variables. It was therefore used to establish the relationship between rainfall and water supply in Bamenda I.

From Table 3, the Pearson's correlation coefficient is 0.701. This positive value shows a direct relationship between monthly rainfall and monthly water supply in Bamenda I. Meaning that an increase in rainfall amount will lead to an increase in quantity of water supply while a fall in rainfall will condition a fall in water supply quantity (in monthly or seasonal bases). This is because an increase in rainfall amount due to the outbreak of the rainy season will progressively recharge the water table and increase its level thereby increasing the volume and level of surface and groundwater, respectively, this will consequently increase quantity of water abstracted and supplied in the municipality.

Though the correlation coefficient indicates a strong positive relationship (+0.701), it is not up to a total positive relationship (+1), implying that other factors influence quantity of water supply in Bamenda I. To determine their contribution, the coefficient of determination was calculated, and the proportion of rainfall was determined. This coefficient gave an R^2 value of 0.49 and a rainfall proportion of 49.14%. This percentage shows the contribution of rainfall on the changes observed in quantity of water supply, therefore the remaining 50.86% indicates the contribution of other factors. These factors include demographic increase, urban expansion, infrastructural, and managerial factors.

Table 3 Pearson's correlation coefficient between monthly rainfall variability and water supply

Variables tested	Pearson's correlation coefficient (r)	Coefficient of determination (R^2)	Proportion of rainfall in the change
Rainfall (mm) and water supply (m^3)	0.701	0.49	49.14%

Source: Author (2019)

High Rainfall Intensity and the Water System

Occurrence of torrential rains in this area is usually accompanied by heavy down-pours, which generally constitutes more of overland flow than infiltration. This is because the steep topography limits chances of infiltrate and rather permits runoff. These conditions will therefore lead to occurrence of surface runoffs, which flow into rivers and increase their volumes. This increase in river volume will increase its force and action within the channel, and this will lead to the occurrence of landslides along its banks. On the other hand, runoffs mostly transport sediments and debris, which ends up blocking water intakes and damaging transmission pipes, thereby preventing water input and output into and out of the system (Plate 1).

As seen on Plate 1, landslides occur due to intense erosive activities along the river course due to intense rainfall episodes (Photo 1). Equally, the transportation of sediments and deposition within the river channel leads to blockage of water collecting chambers and consequently, no inputs into the water system. Also, the occurrence of gully erosion will lead to the exposure of and further damage on pipelines and in such situations, outputting will be hindered, (Photo 3). When these phenomena take place, the entire water system is disrupted, and water crises are witnessed in the municipality.

Droughts and Natural Supply

When meteorological droughts persist, hydrological droughts develop. The term "drought" can have different meaning to different people, depending on how a water deficiency affects them (Moreland et al. 1993). To the people of Bamenda I, droughts are condition of persistent dryness that leads to a reduction in stream flow and groundwater levels, thereby causing water shortages. Prolonged occurrences of this drought therefore lead to shrinking of their water sources (streams, springs, and wells) and in some cases, it leads to the complete disappearance of some streams, (Plate 2).

As seen on Plate 2, there is a drastic fall in the volume of stream flow (Photo 1), shrinking of springs (Photo 2), and complete disappearance in stream flow as seen on Photo 3. These situations are found in Nkar, Ntanche, and Alahnting, respectively.

Photo 1   Photo 2   Photo 3

Plate 1 Impacts of heavy downpour on the water system

Photo 1   Photo 2   Photo 3

Plate 2 Impacts of drought on natural supply

Photo 1   Photo 2

Plate 3 Impacts of drought on supply points

This is caused by three main factors: firstly by lack of precipitations to recharge surface and groundwater, secondly by prolonged evaporation that lead to direct water lost, and lastly by uptake and transpiration from vegetation that reduce both ground- and surface water. Intense insulation and high temperatures often amplify this situation. This result supports the thesis of Adefolalu (1993). On the idea that "climate change can lead to drying up of springs and is projected to reduce surface and ground water resources significantly in most subtropical regions." This reduction in stream flow will reduce quantity of natural supply, thereby reducing input of water to the system, and this will lead to a drastic reduction in total quantity of water supplied within the municipality.

This fall in supply has led to the shutdown of most public supply spots since available water quantity is no longer sufficient to supply all points, (Plate 3).

Plate 3 illustrates shutdown of water supply points due to scarcity in water resource. This is the case of Akumbele and Ntafebuh on Photos 1 and 2, respectively. This occurs because a fall in natural supply due to meteorological drought will lead to a reduction in input to the water system. This reduced in input will lead to a reduction in the quantity of water abstracted and stored in reservoirs, consequently, there will be shutdown of most public supply points since water quantity is no longer enough to freely satisfy the population.

Adaptation Measures

Households and stakeholders have put in place a wide range of measures to adapt to rainfall variability and reduce its effects on water supply in Bamenda I. They range from drilling of boreholes and wells, spring source coverage, use of rainwater collection systems (during rainfall episodes), and through the application of water catchment protection and management measures such as construction of life fences, planting of water friendly trees, slanting and enlarging river banks, and engaging the local population in catchment protection.

Drilling of Boreholes Wells

Surface water sources have become exposed and vulnerable to climatic and environmental hazards such that people were forced to adapt through use of alternative water source. This change was from the use and dependence on surface water sources to groundwater source, which are less exposed and sensitive to the effects of climatic hazards. These surface water sources include boreholes and well.

Spring Source Coverage

To reduce the effects of runoffs, mudflow, and consequent pollution on spring source, the local population constructs protective layers on spring sources. These layers are locally constructed using cement and sand (Plate 4, Photo 2). This is done by joint contribution from the inhabitants of the quarters concerned.

Use of Rainwater Collection System

The use of rainwater over the years has gained its importance as a valuable alternative or supplementary water resource. This is due to the growing scarcity of

Photo 1.   Photo 2.   Photo 3.

Plate 4 Adaptation to climate-induced effects on water sources

ground- and surface water caused by extreme weather conditions. Households therefore witness waters shortage and seizures in the dry season as well as in the rainy season. Seizures in the rainy season are often because of intense rainfall that creates damages around stream catchments preventing water inputs and limiting supply. To remedy water shortage during these periods, some households install rainwater collection systems to increase their water supply. Others use containers to collect water dripping directly from roofs. All these measures are clearly illustrated on Plate 4.

Plate 4 illustrates adaptation measures mainly by the local population to combat the effect of climate extreme on water sources in other to maintain and perhaps increase domestic supply quantity. Photos 1, 2, and 3 shows use of wells, spring source coverage, and rainwater collection system, respectively.

Photo 1 shows a domestic well, and it consists of a digged hole down to the water level. Its water is stored within the hole, and its depth is less compared to that of a borehole. This water source is less vulnerable to climatic extremes, and it ensures a steady supply regardless of seasons. Its steady supply is because it is alimented by aquifers, which are less vulnerable to extreme weather conditions. This water source is therefore used as an adaptive source.

The arrow in Photo 2 points out the protection chamber constructed on a spring outflow. This is to prevent the runoff and mudflow into spring sources. This preventive measure will ensure steady water supply without interruptions especially during and after rainy periods. This situation is common in Ntanche and Aningdoh, respectively. Photo 3 illustrates a household rainwater collection system and this system is used to harvest and store water from roofs during rainfall episodes. Rainwater harvesting and storage is done with pipes, which are connected to gutters permitting to link water from the roof to the storage containers. Water is harvested and stored for further use, especially when the stream supply system fails.

Other domestic measures of adapting to water shortage brought by extreme weather conditions includes: increase use of storage containers, buying mineral water from shops and from water vendors, regulating water use, use of stream water, and practicing of household water treatment and storage measures. Those with borehole at times use generators to pump water into reservoirs during periods of power failures. This is common in dry season due to a fall in dam discharge caused by excess meteorological drought that reduces water recharge in the dams. This reduction in water recharge reduces power supply to the electrical system, and this increases the rate of power failures. As such, those with electrical-based water systems are forced to use alternative power supply measures to electrify their systems. The frequency of implementation of adaptation measures depends on the extent to which rainfall variability affects their sources of supply.

Adaptation Through Catchment Protection and Water Management Measures

Water catchment management (WCM) refers to a range of measures and policies implemented on water catchment in other to reduce environmental, climatic, and anthropogenic actions. It equally involves the management of water catchment through taking precautionary measures to prevent or reduce environmental and climatic hazards on drainage basins and or catchment areas. It also concerns the sustainable use of water resource and land within the watershed. Actors involved in these activities include the Ministry of Water and Energy (MINEE), the Cameroon Water Utility Corporation (Camwater), and the Bamenda I sub divisional council. There is equally the role of the local community and anonymous individuals. These WCM measures are aimed at providing sustainable supply of water in terms of both quantity and quality of water. They include:

- **Construction of life fences.**
  To fight against erosion, there was the planting of life fences round the water catchment areas, both by the subdivisional council and the Camwater. These fences are made of wires and backed up by trees to prevent excess erosion from the surrounding into catchment and river channels. They equally help to prevent grazers from invading the catchment with their cattle. All these measures reduce the rate of damages by erosion on river channel as well as prevent the abuse of water resource by grazer.
- **Planting of water friendly trees.**
  To combat the problem of excess hydrological droughts in Bamenda I, there was a massive planting of water conservation trees (50,000 *Pygeum africanum* tree) around water catchments. This was done by both the Bamenda I subdivisional council and the Bamendakwe Development and Cultural Association (BAMEDCA). This tree has a little quantity of water uptake and disfavors transpiration. It equally has a large canopy that sheds the catchment and prevents evaporation. Its roots equally hold soil particles together, thereby limiting occurrence of landslides in water catchment. On the other hand, there is a strict restriction on the planting of eucalyptus trees around water catchments. This is due to their high water needs and uptake. All these reduce the level of water lost, thereby conserving water within catchments and ensuring sustainable supply.
- **Slanting and enlarging riverbanks.**
  Due to the occurrences of landslides along river courses, there was the need to dig along riverbanks in other to gentle the slopes and reduce the action of gravity, which often bring down landmasses along steep riverbanks. Slanting in most cases widens the riverbed, this reduces both the level and the strength of river flow, and this will therefore reduce the erosive action of the river on its banks. This activity is at times accompanied by the planting of trees along river banks which hold soil particles together and reduce the movement and land masses into river channels. This activity is mostly carried out by Camwater.
- **Engaging local population in catchment protection.**
  This involves giving scientific knowledge to the local population on how the catchment operates, thereby permitting them to protect the water catchment at the local level and ensuring a lasting supply of water in catchments. This includes sensitizing the local population (especially those around the catchment) on sustainable agricultural practices around the catchment, such as the importance of agroforestry and antislope wise cultivation in other to prevent and reduce

excess erosion and deposition of sediments in river channels. It equally concerns placing restrictions on practices such as grazing within water catchment, farming, and deforestation along river courses. It also involves restriction on waste disposal in river channels and around water catchments, since solid and liquid waste disposal in rivers and open space leads to diverse kind of health problems including water and airborne diseases (Achancheng et al. 2003). On the other hand, this is concerned with the massive participation of the entire community around water catchment during activities such as planting of trees and building of life fences. A reduction in these activities and participation of the local population will provide a friendly milieu for conservation of water in the catchments and hence to improve water resources while ensuring the productivity of any water body for the community that depends on it.

Adaptation measures are equally implemented on storage and distribution facilities. This was in a bit to increase water collection, storage, and hence distribution.

Adaptation on Storage Facilities

To address the problem of water lost to the environment by natural springs, amendments were made on spring sources and reservoirs constructed on them to collect and store flowing water. There was equally the construction of water tower to abstract and store water from aquifers, all these in a bit to store water and prepare for periods of high water demand. Doing so equally reduces the direct impacts of climate extremes on springs especially pollution from runoff and mudflow due to intense rains. This was done by the Bamenda I subdivision council and by the Cameroon-China cooperation. These reservoirs were constructed at Menka, Abangoh, and Mendankwe with a carrying capacity of 18 m^3 each.

Adaptation on Distribution Facilities

To limit the damaging action of erosion, floods, and other external factors on pipelines, the water management committee proceeded to the use of metallic pipes on strategic points (in highly vulnerable zones to erosive actions) on the distribution network. There was equally a considerable increase in depth of pipelines especially in areas with loosed soil particles. These were aimed at minimizing damages on pipe network in other to maintain supply. The water management committee equally puts in place distribution trucks (mobile reservoir) to ensure supply especially to extreme quarters on the distribution network. This is mostly common in the dry season were pressure is not enough to pump water to all localities due to low water levels in reservoirs. This service is common in Achichum, which is furthest on the distribution network, and sometimes in Ayaba, which is higher in altitude. In most cases, the trucks are recharged in areas with adequate water supply and distributed in those with limited supply. All these facilities will therefore increase supply in periods of shortages and in vulnerable zone, thereby increasing the total quantity of water supply in Bamenda I. In addition, during situations where available water cannot adequately satisfy the entire population, there is water rationing to ensure equitable supply to all subscribers.

Conclusion and Future Prospects

Rainfall is one of the climate parameters which affect water in the entire drainage basin, thereby leading to consequent effects on rivers and their tributaries. This study was based on the impact of rainfall variability on the quantity of water supply in the Bamenda I municipality. Rainfall amounts in the municipality fluctuate around the normal. Its variability is indicated through fluctuations in interannual rainfall variability and anomalies, and through monthly rainfall variability and anomalies. These anomalies are either positive or negative indicating periods of more or less rainfall amounts. Rainfall variability equally manifests through variations in number of rainy days and through fluctuations in rainfall intensity. The rainfall regime was gotten using climatic indices, and the Pearson's correlation coefficient was conducted to determine how quantity of water supply responds to rainfall variability. However, rainfall conditions seasonal variation in quantity of water supply to an extent, since supply is generally higher in rainy than dry seasons (direct relationship), though some cases occurred where periods of intense rainfall instead leads to a drastic reduction in supply (inverse relationship). Extreme weather events such as heavy rains, flooding, and drought are increasingly becoming common in the municipality. Their occurrence distorts the functioning of both the hydrological cycle and the water system, thereby affecting the quantity of water supply in the municipality.

Future Prospects

Since the study was carried out in the Bamenda I municipality (Bamenda highlands), there is need for further research to expand the geographical scope of the study, in other to do a comparative study of two different drainage basins or water catchments to assess the level of responsiveness of water resources to rainfall variability.

Result from the study showed that beside rainfall variability, there exist other factors, which greatly affect quantity of water supply in Bamenda I. Therefore, further research should be carried out in other to access the validity of these factors.

Equally, an improvement in hydrological research will provide a room for adaptation measures. This is because more research in the domain of water supply will identify more threat to the natural functioning of water catchments (drainage basins), and therefore more suggestion will come up to improve adaptation measures on water catchment.

References

Achancheng E, Lydie SA, Temitope DT, Corntin YC (2003) Urban cities and waste generation in developing countries: a GIS evaluation of two cities in Burkina Faso

Adefolalu (1993) Precipitation, evapotranspiration and the ecological zones in Nigeria

Dhaka. http://www.wateraid.org/documents/plugin_documents/060721_tubewell_guidelines.pdf

Flyer, Bamenda I Council (2018) Physical and human environment of Bamenda 1.

http://www.google.com/search?q=pangire+date+of+creation&client=ucweb-b&channel=sb

http://www.who.int/water_sanitation_health/publications/vision_2030_summary_policy_implicati ons.pdf

Hayward O, Mata LJ, Arnell NW, Döll P, Kabat P, Jiménez B (1987) Climate and Chlorophyll: Long-term trends in the central North Pacific Ocean

IPCC (2008) Climate change and water. Intergovernmental Panel on Climate Change Working Group II technical support. Jewitt, G.P.W, Schulze R.E (1999). Verification of the ACRU model for forest hydrology applications. Water SA 25:483–489

IPCC (2014) Managing the risks of extreme events and disasters to advance climate change adaptation. A special report of Working Groups I and II of the intergovernmental panel on climate change. Cambridge University Press, Cambridge

Jiduana GG, Dabi DD, Dia RZ (2011) The effects of climate change on water and agricultural activities in selected settlements in the Sudano-Sahelian Region of Nigeria. Arch Appl Sci Res 3:154–165

Mark E, Andrew A, Joseph L, Jamie B (2008) HWTS education: a hidden success in emergency situations" Presentation at the "Water and Health: Where Science Meets Policy" conference. Chapel Hill, USA. October 26, 2010

McKee TB, Giddingset, Nolan J (1993) Standard precipitation index (SPI) and methods of accessing rainfall variability. The Relationship of Drought Frequency and Duration to Time Scales. 8th Conference on Applied Climatology, Anaheim, 17–22 January 1993

Moreland (1993), Burdon DJ (1985). Groundwater against drought in Africa. In Hydrogeology in the Servic of Man, Mémoires of the 18th Congress of the International Association of Hydrogeologists. Cambridge. http://iahs.info/redbooks/a154/iahs_154_02_0076.pdf Accessed 11 November 2010

PANGIRE (2013) Plan D'Action National de Gestion Intégrée des Ressources en Eau: Etat des lieux du secteur

Parry ML, Palutik JP, Van Der Lind PJ (2007) Water supply sanitation (WSS) in Moldova, climate change: impacts, adaptation and vulnerability. Contribution of working group II to the Inter-governmental Panel on Climate Change. Cambridge University Press, Cambridge

Tsalefac M (2007) Variabilité climatique, crise économique et dynamique des milieux agraire sur les Haute terres de l'ouest Cameroon. Thèse de Doctorat d'Etat de lettre et the science humaine, spécialité, option climatologie, Université de Yaoundé, 564p

Tsalefac M, Ngoufor R (2007) Afrique Centrale, le Cameroun et les changements globaux

Walsh R.P.D, Lawler D.M (1981) Methods of determining seasonality index and methods of accessing rainfall regimes. Rainfall seasonality: Description, spatial patterns and change through time

Wilby RL (2006) Van Vliet and Zwolsman (2008) A framework for assessing uncertainties in climate change impacts, low-flow scenarios for river Thames, UK Sustainable land and water management policies and practices: a pathway to environmental sustainability in large irrigation systems. Land Degrad Dev 19:469–487

12

Clean Energy Technology for the Mitigation of Climate Change: African Traditional Myth

Abel Ehimen Airoboman, Patience Ose Airoboman and
Felix Ayemere Airoboman

Contents

Abstract

The global Anticipated Energy Transition Period (AETP) is one that all stake-holders must embrace with respect to curbing energy poverty, thereby addressing issues related to climate change especially in the sub-Saharan region of Africa. The region is endowed with abundant richer, cleaner, and affordable energy sources, majority of which has remained untapped due to many reasons, one of which is tied to the socio-cultural traditional beliefs and value systems of the

A. E. Airoboman (✉)
Department of Electrical/Electronic Engineering, Nigerian Defence Academy, Kaduna, Nigeria
e-mail: airobomanabel@nda.edu.ng

P. O. Airoboman
Department of Biotechnology, Nigerian Defence Academy, Kaduna, Nigeria

F. A. Airoboman
Faculty of Arts, Department of Philosophy, University of Benin, Benin City, Nigeria
e-mail: felix.airoboman@uniben.edu

citizens. This has forced majority of the inhabitants to continue to rely on the use of non-biodegradable materials for the purpose of cooking and many other activities. This value system, therefore, contributes to have had an adverse effect on the climate and also on the health of the citizens most of whom are women and children residing in rural areas. The outlook on the AETP, their effect on climate change, the use of Clean Energy Technology (CET) domestically, the various strata expected to come with the AETP, the socio-cultural dynamics in terms of acceptability by all (rural, peri-urban, and urban areas) is addressed in this chapter. The chapter concluded by designing a CET model that could assist in planning for the AETP and mitigating climate change.

Keywords

AET · Africa · CCS · CET · Biofuel · Environment · Morality

Introduction

Energy plays a vital role in socioeconomic affairs of people, but it must be available in a desired quality and quantity to be able to meet their needs. Only 290 million people out of the 915 million people in sub-Saharan Africa has access to electricity (IEA 2014; Aboua and Toure 2018). It is equally expected that in the years to come, the population in Africa will grow rapidly, and this may cause a major challenge to the international community due to the slow energy penetration it has received over the years. One of the goals of the appropriate authorities in years to come would be to phase out the use of coal and oil in the market due to their harmful effects on the environment, livestock, and people and then replace them with clean energy for residential, commercial, and industrial uses. According to IRENA (2019), the price of solar PV, solar battery, and solar wind has reduced by 70%, 40%, and 25%, respectively, since 2010, and this is an indication that the Clean Energy Technology (CET) is gradually penetrating the market due to reduction in prices and an improvement in the technology. Generally, energy consumption has direct effects on the role of individuals and communities, whether local, urban, or peri-urban. The idea is that energy contributes to the technological, sociocultural, agricultural, educational, as well as psychological standards in a given settlement; thus, it should be accessible, reliable, affordable, and available when needed. One of the technological penetrations as a result of Anticipated Energy Penetration (AEP) would be the use of Clean Energy Stove (CES) for cooking. Energy comes in different forms from either biomass or biogas. Biomass, as an energy source, has been appreciated by people over the years because it is readily available and appears to be cheap. However, it is anticipated by the United Nations that by the year 2040, electricity should be appreciated by people all over the globe as their first choice of energy. This chapter, therefore, looks at the CET during the pre-AET and post-AET period, their effects, and how they would contribute to the sociocultural dynamics of settlements in sub-Saharan Africa and the sustainability of their environment.

Clean Cooking Energy and Their Challenges

According to WEO (2017), 2.8 billion people across the globe is presently denied access to CET. Majority of them presently reside in Asia and Africa. Similarly, the World Health Organization (WHO) recorded that 7.7% of death rate worldwide is the result of indoor pollution and incomplete combustion; and mostly, women and children are the most vulnerable. One of the contributing factors among others is the inefficient manner in which the solid biomass fuel is used. Furthermore WEO (2017) also affirmed that four out of every five persons in sub-Saharan Africa are yet to appreciate the CET; rather, they prefer the use of solid biomass as fuel, and the effect of this is deforestation, desertization, pollution, erosion, loss of species, global warming, impaired human development, and scarcity of raw materials for other technological purposes. During the AET period, it is expected that solid biomass fuel such as firewood, agricultural residue, coal, animal dung, and generic organic waste would be phased out gradually and replaced with electricity. It is also envisaged that during this period, the choices for other forms of energy other than electricity available across the globe should be reduced drastically because the more the choices are available, the greater the burden in transiting to Clean Cooking Energy (CCE). The advent of the AETP is expected to gradually integrate the CET into the market and this technology will be geared to address inefficiency in cooking methods while speeding up the rate of clean energy penetration and otherwise. However, for a smooth transition to be recorded, reviews are needed due to availability, adequacy, stagnant growth as well as political reasons. A novel method to CCE should therefore take cognizance of various dimensions and complex nature owing to the stringent measure in eliminating localized ways of cooking. For the locals, this is already a way of life, a doctrine, a culture and a belief system.

Before this transformation, there is the need for moral requirement to properly educate and enlighten the indigenous people to ensure they are able to make informed consent and autonomous decision without any imposition, coercion, manipulation, or persuasion and the people's autonomy is respected (Faden and Beauchamp 1986; Beauchamp and Childress 1994). Besides, these technologies should be friendlier to the people and their natural environment; they should not make the environment hazardous and uncongenial to the people. They should conform to ethical duty (Desjardins 2006: Rachels 2007) and the utilitarian end (Rachels 2007) of the people and their environment, both at present and in the future. They should not be adopted on the sole aim of profit making and on the motive of instrumentalization of the people and their environment. Relying on the Divine Command Theory of Morality (DCTM), one is prompted to ask whether God would require, permit, or prohibit people being deceived, cajoled, or forced to jettison a way of life that they have been accustomed to and lived by or force them to accept a view and way of life that they may consider inimical to their well-being (Desjardins 2006). Unless these fears are allayed, there is no guarantee, for example, that when the Clean Cooking Stove (CCS) and other new technologies are provided to the locals during the AET period, they would not be stacked. Depending on the temperament of the people, there is the tendency to reject or put up resistance when new ideas and ways of life are introduced to people who already feel secured in their comfort zones, unless adequate care is taken. There is the tendency on trado-cultural ground on the part of traditional rulers or chiefs and traditional doctors to have their meals cooked in special dishes and specified manners or methods and styles, if they are to retain their authorities, powers, efficacies, and sanctities, which may not be amenable to newly introduced alternative technologies. There may also be difficulty of embracing and adapting to new ways of life. Adjustments may cause lots of inconveniences and discomfort to the extent that the people may want to revert to their accustomed and secured styles of living. These may cause some friction in accepting newly introduced technologies.

Another issue of concern is that it pleases the locals to walk some kilometers to gather firewood at no cost rather than trekking some meters to expend their money in buying energy at a high price. Infact, some may not have money to expend in purchasing alternative energy due to their degree of poverty. It would be morally wrong to sway people away and "coerce" them into accepting a way of living they are unaccustomed to without adequate enlightment programme and also without providing them with an affordable, safe, and simple alternative especially, if their previous methods and modes of living are innocuous, that is, it constitutes no harm or danger. In this way, the enforcing government, agency, or any other body fails in their moral duty.

For instance, in Nigeria, cooking in urban and rural areas is actualized by 50% and 99.8% of nonbiodegradable fuels, respectively. The high margin recorded in the rural areas is because they are unable to pay for a neater and cleaner substitute, lack awareness, and lack value systems in addition to the lack of accessibility to such fuels. Another challenge faced by the rural areas is stacking of any technology on the ground that they are scared of operating any electrical gadgets. Although their present technology might look cheap in the interim, other cooking options would be cheaper in the long run (National Energy Policy 2013). If remedial actions are not taken effectively, then the nation's 15 million hectares of forest according to National Energy Policy (2015) could be depleted within the next 50 years, and the goal of AETP might not be met from the Nigerian perspective.

Also, the nonbiodegradable fuel source in Nigeria contributes about 37% of the total energy demand at a consumption rate of $43.4 * 10^9$ kg of fuelwood annually, making an average of 0.5–1 kg of fuelwood per person on a daily basis. This practice, has therefore encouraged the cutting down of trees, thereby making it a lucrative business owing to the huge and increasing demand of fuelwood, thus making Nigeria one of the largest fuelwood markets in the world (EIA 2014).

The cooking energy demand can therefore be represented mathematically in (1) as presented in NECAL (2015):

$$CED = \frac{CD}{H_H} \times \sum_{i=1}^{n} H_{H_i} \qquad (1)$$

CED = cooking energy demand, CD = cooking demand, H_H = household.

According to (ECN 2015), 71.9 million hectares of land in Nigeria is fertile, therefore making it a good potential for biofuel production other than fossil fuel. However, with the level of hunger and poverty, low gross domestic product (GDP), etc., it might be difficult to sacrifice the primary sources of biofuels which are actually foodstuff like cassava, sugarcane, soya bean, oil palm, etc. for the purpose of generating clean energy for domestic purposes. Although (Kela et al. 2015) have shown the introduction of CCS which includes clay-based improved stove, energy-saving stove, various solar cookers, double-pot improved stove, etc., their use is still at the experimental stage and therefore requires financial support to increase the level of awareness as well as mass production.

As a way forward, Olaniyan et al. (2018) proposed the use of nonfood crops for the production of bioenergy in some West African countries. However, based on the arable hectares of land in Nigeria and other African countries, another way forward is the genetic modification of certain biofuel plants for the purpose of producing fuel directly for domestic use and for the purpose of cooking. Plants with high calories can also be modified to produce more sugar than expected which can be used for fuel production. From these, excess sugar can actually be picked from the plant for the purpose of fuel production, and domestic use while the other part of the plants continues to remain edible for consumption thereby guaranteeing food security. Other than this, with respect to the AET, the full dependence on electricity for domestic use is also anticipated. In Fig. 1, the stages of AET and CET period are described. In Figs. 2 and 3, the level of merchandising of fuelwood and the use of solid biomass in cooking at restaurants in urban area are presented. From Fig. 3, it is shown that any deviation from this mode of cooking implies that the traditional smoky flavor is absent, hence reducing customers' patronage in the restaurant.

Role of Government During the AET Period

There is a call across the globe for nations to move from "oil-driven economy" to "clean energy economy." Various changes are expected to be noticeable worldwide during this period with respect to regulations, policies, financing, consumption pattern, and so on. Now, what will be the fate of the individual citizens of those countries whose economy is strongly dependent on oil if the issues that directly affect them, and if what their daily living depend upon, are left at the hands of their decision-makers and the international community? Hence moral rectitude is required in convincing indigenous people. Again, given the abundant resources in their country, would they also transition during the AET period? If yes, at what pace? If no, why? The world is fully aware of the abundant primary sources of energy in Africa which are yet to be harnessed, and there is a call for a carbon-free environment in the nearest future, yet Africa continues to be the largest market for generator dealers across the world due to the high level of unreliability in the electric power sector. Part of the effects of generator use is that it constitutes health hazards to humans and the natural environment. It contributes to acid precipitation, depletion of ozone layers, carbon poisoning, and so on (Airoboman and Tyo 2018).

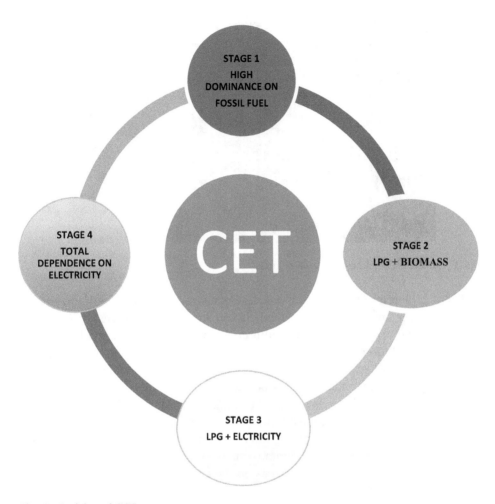

Fig. 1 Anticipated CCS stages

Fig. 2 Merchandising on fossil fuel in North-West Nigeria

Fig. 3 Commercial cooking with solid biomass

There is therefore the need for stakeholders across the globe to deliberate on this issue to see how this energy can be managed in the most efficient way so that it can continue to be a means of wealth creation for the nations involved rather than asking or "forcing" them to give up their natural resources and go green. Also, if sustainability is to be recorded, then the number of portable generator sets penetrating the African market must be reduced and replaced with a more eco-friendly one. The role of each respective authority during the period would encompass awareness programs through the use of various advertisements, signs, agencies, and jingles, among others, to inform their citizens on the need to transiting from the use of solid biomass fuels to CET. The appropriate authority is also expected to enlighten its citizens on the cancerous effects that these fuels would cause in the long run and would continue to cause if their use is not abolished or optimized. The distance traveled by women and children to gather firewood contributes to the impairment of human development. Besides, the time wasted in all these processes which could have been valuably geared toward other meaningful engagements is also a call for the appropriate authority to answer. Climate change and depletion of the ozone layer is another strong point that the government is expected to use in convincing its citizens. How well this will sink into the hearts of the people is another cause of worry for the government. Therefore, integration of stakeholders and nongovernmental organizations (NGOs) whose job would be to interact and connect with the people one on one; study their beliefs, cultures, and ways of life; and then finally come out with the best and sustainable CET for a given community. It is also expected that electricity tariff should be made very simple and affordable. There should be guarantee by the concerned authority to its citizens that electricity, which is now their first choice of energy, will always be reliable and available when needed. Furthermore, the prices of the CCS should be subsidized and made at various sizes so that even the poorest person in a rural, peri-urban, and urban area can afford them.

The Psychological Myth of Locals

Universal access to energy especially in the local community continues to be a worldwide priority by the year 2030, there is a need to look at how the lack of access to energy has contributed to development impairment and most especially how the CET would be appreciated by locals and the entire citizens of a given country especially in Africa. A key factor to be considered here is that the traditional fire does not only serve as a cooking stove, but it also plays other significant roles such as providing warmth during cold, drying of clay object, preservation of some agricultural product till the next planting season, prevention of weevils from some harvested crops, chasing out of wild animals from a vicinity, heating of the environment during a particular time of the year, a source of light during moonlight stories, waste disposal, drying of clothes, security purposes, and a sign of faith as practiced by some orthodox churches during a certain time of the year. If CET could penetrate the African continent, then it is expected that it should be able to address all these "ways of life" in the simplest manner and with simple technology. Besides, the CET should be cost-effective. The idea behind the use of CCS is simplicity in size and cost. If this is not achieved, various options to fuel available could make the CET objective a herculean task. Furthermore, there are some unique cultural preferences that might be difficult for locals to give up, and one of such examples is the traditional smoky flavor on meals prepared. The locals believed strongly that such flavors and taste could only be gotten through the use of fuelwoods and coal, and as such, the use of CCS could change the taste of their food. Even the rich often times visits small canteens where the source of energy used is firewood in order to get that traditional smoky flavor which makes their food sweet rather than visiting big restaurants whose source of energy is mostly microwaves because of the fear of radiation that causes cancer. The irony that the locals prefer to prepare their food using the traditional firewood because of its "simplicity" rather than using any other sources such as modern energy stoves because of their "complex nature" is also a call for concern and worry.

Simplicity for them is a priority and it comes before availability. Another issue of concern is that it pleases the locals to walk some kilometers to gather firewood free of charge rather than trekking some meters to expend their money in buying energy at a high price. In addition, some may not have this money to expend at all to purchase alternative energy due to their degree of poverty. Despite this "way of life" (which we may call archaic and uncivilized in the modern times), it is on record that people in villages live longer than people in the city. Hence, selling a technology from the city to the villages requires stringent measures. It would be immoral to impose, coerce, or deceive the locals to accept a method or an alternative way of life without informed consent, free choice, or adequate enlightenment, including relevant and adequate knowledge about the benefits and harms of the technology to be introduced. Concerning what makes an issue moral, Barcalow (1994) argued among others that the choices people make about such issues affect their well-being and the well-being of others; and such well-being may be physical or psychological. Such actions should also be capable of causing benefit or harm. It would also be a moral dilemma for those knowledgeable or well informed about the actual and potential hazards of the habits and practices of a people to allow them to continue in those habits whether or not it directly affects them, when they are unwilling to change.

A Look at Environmental Ethics

The heavy burden imposed on nature in recent times because of various activities by humans, it is now impossible for nature to regenerate itself the way it used to. Environmental ethics could therefore be put forward as one of the ways of tackling environmental issues and simultaneously meeting human needs. According to Omonzejele et al. (2017), every culture possesses certain notion of what is good and bad and what "ought" and "ought not" to be done. Therefore, the bad aspects of using fossil fuel plants for the purpose of heating, cooking, drying, security, etc. by rural areas may be left hanging as it is already a value system and a way of life; hence, all they see is the "good" aspect of using fossil fuel as the word "bad" may sound offensive because of what it projects and represents. The rural dwellers are of the opinion that the human life has value. Then, if prevented to sufficiently interact with its environment, it may be tantamount to saying the value of the non-human life is superior to the human life, and this could negate their traditional and cultural doctrines of Anthropocentrism. Therefore, a point of alignment between the non-human life and the human life needs to be defined to address these stringent issues with respect to educating the rural dwellers especially on the doctrine of Sentientism which opined that an element of consideration should be afforded to the non-human life. Also, the doctrine of Biocentrism strongly stated that all life is sacred! In this sense, the human life must refrain from cutting branches from trees for any purpose. The doctrine of Ecocentrism also asserts that nature is the center for the existence of human life upon which this life is parasitic. In Airoboman (2017), one of the reasons for environmental concern is to curb issues and proffer solutions on issues relating to climate change. The author stressed that development is a function of the level of interaction with the environment. However, with the present climate change challenges, one may ask if the gain as a result of development is worth the trouble posed by the climate? If yes, should the underdeveloped nation aspire to develop or remain static developmentally because of possible fear of climate-related issues? These questions are begging for answers. Therefore, environmental ethics should be included in high school and colleges curricula in order for all parties involved to be guided accordingly.

In Fig. 4, the occupant of the apartment uses LPG for cooking. However, the color of the ceiling shows otherwise, and this may be a result of the inefficient way the LPG was used, thereby constituting environmental challenges as well health challenges especially to the occupant of the apartment. CET should not just be about the use of clean energy but also of the way(s) it is used with respect to each environment. As a means of addressing the myth behind cooking using the CET, the authority concerned would have to convince the masses that:

Fig. 4 A polluted kitchen

1. The technology behind CCS would be very simple such that it would be easy to operate.
2. It is not time consuming because the energy required would always be available rather than traveling longer distances to fetch firewood.
3. Maintenance of the CCS would also be made very simple and available.
4. The locals would be involved directly in the maintenance, thereby creating jobs for them.
5. It is the presence of carbon black that affects visibility. Hence, there should be a paradigm shift.
6. There are negative impacts on the method they have been accustomed to have on the natural environment.
7. There are actual and potential hazards, which impact the human beings, future generations, and general ecosystem.

In addition to the above, there is also a need by the authority concerned to introduce CCS designed to give that traditional smoky flavor in food, as well as addressing the fear in operating CCS because of its "complexity" and fire hazards.

The Way Forward

As the world continues to strive in meeting its objectives regarding universal access to energy and CET, there is a need to deplore methods to be followed in order for these objectives to be met. In Fig. 5, a model which incorporates planning, policy making, regulations, health, and safety and standards is proposed. In this scenario, analysis is done at various levels to get a master plan for CET in the near future.

1. The planning level involves various coordinated policies and strategies to be adopted for a smooth AET.
2. The policy level should incorporate the drafting and implementations of policies.
3. At the regulation phase, it is expected that suitable regulation standards both technical and economical are taken into cognizance. The direct involvement of government is expected here because it has to protect the consumers' right and ensure that any tariff adopted must be simple and cost-effective.
4. Health and environment protection agency regulates the impact of the technology on the health and safety of the environment as well as the personnel.
5. At the standard phase, regulations such as conformity to WHO standards should be enforced. It should be noted that it is not only laboratory test that is required here; other regulations that will ensure sustainability through availability of spare parts and maintenance must be taken into cognizance.
6. The CET room investigates the report from various agencies, regulators, and policy makers to see the potentiality of the CET penetration in a given area, and when satisfied, they move to the next phase as shown in Fig. 6.

Fig. 5 CET policies and regulatory phase

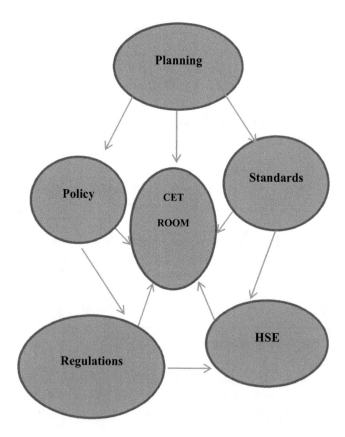

Fig. 6 CET implementation
stage

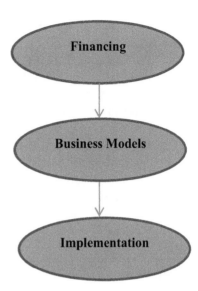

The second stage of the model comprises the financing, the business models to be adopted and finally the implementation.

1. At the financing phase, it is expected that there is enough fund to support investors, willing to go into the CET business. There should also be a body under the CET that manages and gives out funds to potential investors. There could also be a Public Private Partnership that allows these investors to build, operate and transfer the technology at the required time.
2. Furthermore, it is expected that stakeholders in the sector will have to meet and develop business models that favor a given region with respect to the natural resources available in that region.
3. Since the goal of every business is to get return on investment, it is expected that this phase should be time consuming owing to various scenarios analysis and feasibility study needed to ensure a very simple but affordable tariff expected by potentials customers. When all this is put in place, the implementation stage comes to fore. This involves the coordination and supervisory roles needed during installation of CET within a given area. Such supervisory role should be handled by an expert in the field who will ensure that all equipment deploy to the field conform to appropriate standard, pass the LAB tests and also meets the WHO requirements.

Conclusion

In this chapter the injustices that lack of energy access has caused globally and especially in sub-Saharan Africa is highlighted. Energy poverty has slowed down technological infrastructure and has led to a slow economic growth rate. As the world is expected to transition to renewables and CET in the nearest future, it is expected that there would be a swift shift from the use of solid biomass fuel to electricity. Therefore, there is a need for Africans and especially sub-Saharan African to start preparation for the AET in earnest in order not to be caught in the web of uncertainty. With proper preparation, good policies, regulations, proper planning, nice business models can be put in place in the African continent because Africa has been designed to rule the world in the nearest future. Hence we must get our technology right; and in doing this, there must be a universal access to energy especially to women and children. The study also concurrently looks at some socio-cultural and economic impediments that may have impeded this goal, as well as some moral issues involved and attempt to address them.

References

Aboua CA, Toure YFE (2018) Efficiency in West Africa economies: implication for sustainable energy. Ecowas Sustain Energy J 1:65–91

Airoboman F (2017) A survey of environmental ethics: "the substance of ethics and morality" Published by The Department of Philosophy, Faculty of Arts University of Benin, Nigeria, pp 149–169

Airoboman AE, Tyo TM (2018) Operation save energy: the grid defence mechanism in sub-Saharan Africa. Am J Eng Res 7(5):130–135

Barcalow E (1994) Moral philosophy: theory and issues. Wadsworth Publishing Company, Belmont

Beauchamp TL, Childress JF (1994) Principles of biomedical ethics, 4th edn. Oxford University Press, Oxford

Desjardins JR (2006) Environmental ethics: an introduction to environmental philosophy, 4th edn. Thomson Wadsworth, Belmont

ECN (2015) Biofuels Training Manual, Federal Ministry of Energy, Nigeria

Energy Information Administration (EIA) (2014) United States of America. https://www.eia.gov/countries/country-data.cfm

Faden RR, Beauchamp TL (1986) A history and theory of informed consent. Oxford University Press, Oxford

IEA (2014) Africa Energy Outlook: A Focus on Energy Prospects in Sub-Saharan Africa (World Energy Outlook Special Report), pp 1–242

IRENA (2019) Future of solar photovoltaic: development, investment, technology, grid integration and socio-economic aspects, pp 1–73

Kela R, Tijjani A, Usman K (2015) Current status of research and development activities on efficient cookstoves in Nigeria. J Energy Policy Res Dev 1(2):50–62

National Energy Policy (2013) Energy Commission of Nigeria, Draft Revised Edition

National Energy Policy (2015) Energy Commission of Nigeria, Draft Revised Edition

Nigeria Energy Calculator (2015) Energy Commission of Nigeria. Department of Energy & Climate Change (Federal Ministry of Science & Technology). https://www.nigeria-energy-calculator.org

Olaniyan OF, Taal P, Kara AA, Adewale A, Ceesay K (2018) Factors influencing the use of non-food agricultural biomass for renewable energy generation: a stakeholders' analysis in the Gambia, West Africa. Ecowas Sustain Energy J 1:93–104

Omonzejele PF Ukagba GU, Asekhaumo AA (2017) African traditional ethics: the substance of ethics and morality" Published by The Department of Philosophy, Faculty of Arts University of Benin, Nigeria, pp 113–124

Rachels J (2007) A short introduction to moral philosophy. In: Rachels J, Rachels S (eds) The right thing to do: basic readings in moral philosophy, 4th edn. The McGraw-Hill Companies, Inc., Boston, pp 1–19

WEO, 2017. Special Report: Energy Access Outlook – Analysis. https://www.iea.org. Access 14 Mar 2020 15:15hrs

Permissions

List of Contributors

Hillary Dumba
Institute of Education, College of Education Studies, University of Cape Coast, Cape Coast, Ghana

Jones Abrefa Danquah
Department of Geography and Regional Planning, Faculty of Social Sciences, College of Humanities and Legal Studies, University of Cape Coast, Cape Coast, Ghana

Ari Pappinen
School of Forest Sciences, Faculty of Science and Forestry, University of Eastern Finland, Joensuu, Finland

Zakou Amadou
Faculty of Agricultural Sciences, Department of Rural Economics and Sociology, Tahoua University, Tahoua, Niger

R. Mandumbu
Crop Science Department, Bindura University of Science Education, Bindura, Zimbabwe

C. Nyawenze
Cotton Company of Zimbabwe, Harare, Zimbabwe

J. T. Rugare
Department of Crop Science, University of Zimbabwe, Harare, Zimbabwe

G. Nyamadzawo
Department of Environmental Science, Bindura University of Science Education, Bindura, Zimbabwe

C. Parwada
Department of Horticulture, Women's University in Africa, Harare, Zimbabwe

H. Tibugari
Department of Plant and Soil Sciences, Gwanda State University, Gwanda, Zimbabwe

Abdellatif Ahbari and Laila Stour
Laboratory of Process Engineering and Environment, Faculty of Sciences and Techniques, Hassan II University of Casablanca, Mohammedia, Morocco

Ali Agoumi
Laboratory of Civil Engineering, Hydraulic, Environment and Climate Change, Hassania School of Public Works, Casablanca, Morocco

Daniel M. Nzengya
Department of Social Sciences, St Paul's University, Limuru, Kenya

Paul Maina Mwari and Chrocosiscus Njeru
Faculty of Social Sciences, St Paul's University, Limuru, Kenya

Adetunji Oroye Iyiola-Tunji
National Agricultural Extension and Research Liaison Services, Ahmadu Bello University, Zaria, Nigeria

James Ijampy Adamu
Nigerian Meteorological Agency, Abuja, Nigeria

Paul Apagu John
Department of Animal Science, Ahmadu Bello University, Zaria, Nigeria

Idris Muniru
Department of Biomedical Engineering, Faculty of Engineering and Technology, University of Ilorin, Ilorin, Nigeria

Cryton Zazu
Environmental Learning Research Centre, Rhodes University, Grahamstown, South Africa

Anri Manderson
Hoedspruit Hub, Hoedspruit, South Africa

Becky Nancy Aloo
Department of Sustainable Agriculture and Biodiversity Conservation, Nelson Mandela African Institution of Science and Technology, Arusha, Tanzania
Department of Biological Sciences, University of Eldoret, Eldoret, Kenya

Ernest Rashid Mbega
Department of Sustainable Agriculture and Biodiversity Conservation, Nelson Mandela African
Institution of Science and Technology, Arusha, Tanzania

Billy Amendi Makumba
Department of Biological Sciences, Moi University, Eldoret, Kenya

Chukwuma Otum Ume
Agricultural and Environmental Policy Department, Justus Liebig University Giessen, Giessen, Germany

Patience Ifeyinwa Opata
Department of Agricultural Economics, University of Nigeria, Nsukka, Nigeria

Anthony Nwa Jesus Onyekuru
Resource and Environmental Policy Research Centre, Department of Agricultural Economics, University of Nigeria, Nsukka, Nigeria

Royford Magiri and Paul Iji
Fiji National University, College of Agriculture, Fisheries and Forestry, Suva, Fiji

Kaampwe Muzandu
School of Veterinary Medicine, University of Zambia, Lusaka, Zambia

George Gitau
Faculty of Veterinary Medicine, University of Nairobi, Nairobi, Kenya

Kennedy Choongo
Fiji National University, College of Agriculture, Fisheries and Forestry, Suva, Fiji
School of Veterinary Medicine, University of Zambia, Lusaka, Zambia

Zoyem Tedonfack Sedrique and Julius Tata Nfor
Department of Geography, Planning and Environment, University of Dschang, Dschang, Cameroon

Abel Ehimen Airoboman
Department of Electrical/Electronic Engineering, Nigerian Defence Academy, Kaduna, Nigeria

Patience Ose Airoboman
Department of Biotechnology, Nigerian Defence Academy, Kaduna, Nigeria

Felix Ayemere Airoboman
Faculty of Arts, Department of Philosophy, University of Benin, Benin City, Nigeria

Index

Printed in the USA
CPSIA information can be obtained
at www.ICGtesting.com
JSHW061053121023
49903JS00030B/98